Composing Resonance

New Media Theory

Series Editor, Byron Hawk

The New Media Theory series investigates both media and new media as complex rhetorical ecologies. The merger of media and new media creates a global public sphere that is changing the ways we work, play, write, teach, think, and connect. Because these ecologies operate through evolving arrangements, theories of new media have yet to establish a rhetorical and theoretical paradigm that fully articulates this emerging digital life.

The series includes books that deploy rhetorical, social, cultural, political, textual, aesthetic, and material theories in order to articulate moments of mediation that compose these contemporary media ecologies. Such works typically bring rhetorical and critical theories to bear on media and new media in ways that elaborate on a burgeoning post-disciplinary "material turn" as one further development of the linguistic and social turns that have already influenced scholarly work across the humanities.

Books in the Series

Composing Resonance: Culture and Collaboration in Sonic Practices by Mary Hocks and Michelle Comstock (2024)

Networked Humanities: Within and Without the University, edited by Jeff Rice and Brian McNely (2018)

Suasive Iterations: Rhetoric, Writing, and Physical Computing by David M. Rieder (2017)

Writing Posthumanism, Posthuman Writing, edited by Sidney I. Dobrin (2015)

Ready to Wear: A Rhetoric of Wearable Computers and Reality-Shifting Media by Isabel Pedersen (2013)

Mics, Cameras, Symbolic Action: Audio-Visual Rhetoric for Writing Teachers, by Bump Halbritter (2013). *Computers and Composition* Best Book Award 2014.

The Available Means of Persuasion: Mapping a Theory and Pedagogy of Multimodal Public Rhetoric, by David M. Sheridan, Jim Ridolfo, and Anthony J. Michel (2012)

Avatar Emergency by Gregory L. Ulmer (2012)

New Media/New Methods: The Academic Turn from Literacy to Electracy, edited by Jeff Rice and Marcel O'Gorman (2008)

The Two Virtuals: New Media and Composition, by Alexander Reid (2007). Honorable Mention, W. Ross Winterowd/*JAC* Award for Best Book in Composition Theory (2007)

COMPOSING RESONANCE

CULTURE AND COLLABORATION IN SONIC PRACTICES

Michelle Comstock and Mary Hocks

Parlor Press
Anderson, South Carolina
www.parlorpress.com

Parlor Press LLC, Anderson, South Carolina, USA

Printed in the United States of America on acid-free paper.

S A N: 2 5 4 - 8 8 7 9

Library of Congress Cataloging-in-Publication Data on File

2 3 4 5

978-1-64317-454-9 (paperback)
978-1-64317-455-6 (PDF)
978-1-64317-456-3 (EPUB)

Interior and cover design by David Blakesley.
Cover image by Karen Hellyer, using WOMBO's AI "Dream." Used by permission.

Parlor Press, LLC is an independent publisher of scholarly and trade titles in print and multimedia formats. This book is available in print and ebook formats from Parlor Press on the World Wide Web at www.parlorpress.com or through online and brick-and-mortar bookstores. For submission information or to find out about Parlor Press publications, write to Parlor Press, 3015 Brackenberry Drive, Anderson, South Carolina, 29621, or email editor@parlorpress.com.

Contents

Acknowledgments

We are thankful for all the friends, family, and colleagues who have supported us over many years as we developed our thinking and research for this book. We are particularly grateful to David Blakesley, Byron Hawk, and the anonymous peer reviewers with Parlor Press for helping us make our work better. Everyone needs an editor as generous and detailed as Byron Hawk! Many thanks also to Fran Chapman for her excellent copyedits. We want to thank our students at Georgia State University, Atlanta, and the University of Colorado, Denver, who continually encourage us with their willingness to experiment and to inspire us with their compositions featuring sound. This book would not have existed without our students.

Mary is grateful to the published musicians and singer-songwriters whose original music, on-going instruction, and talented performances have supported collaborative work and helped build her musical communities. Many thanks to friends and colleagues who gave permission to document their events, responded to versions of this research along the way, and whose thinking and artistic feedback helped us immeasurably.

Mary thanks the friends and family who helped keep her going for well over ten years while writing this book: Lynn and TL; Ruth, Elaine, and Ms. Pat; Eileen and Jean; Megan and Kally; and sister Beckie. Their kindness and encouragement have meant everything. Mary thanks her family and especially her dad Richard Hocks for his support. She dedicates this book to her mom Mary Elaine Dowling Hocks, who was a veteran writing center director who offered unfailing enthusiasm for Mary's work and her music.

Michelle is grateful to the many colleagues and friends who provided vital feedback at key moments during the writing of this manuscript. Sarah Fields, Sarah Hagelin, Amy Hasinoff, Rodney Herring, Elizabeth Kleinfeld, Lucy McGuffey, Gillian Silverman, Sarah Tyson, and Margaret Woodhull read various versions of these chapters and helped make them substantially better. Thanks also to the Re/Call artists, volunteers, and workshop participants who provided pedagogical and artistic feedback on the sonic practices and heuristics in chapters two and four. A special thanks to Re/Call organizers Mary Caulkins and Louise Martorano for creating a truly collaborative process and space from the very beginning.

Michelle would also like to thank her family. Her mother and father, who moved into her home during the pandemic, provided daily support and encouragement. Her daughter, Sarah, who was often home learning and writing alongside her, helped her keep up the pace and was a terrific sounding board. Finally, she thanks her partner, Suzanne, who has listened to her talk about this project for ten-plus years and has always responded with insight, humor, and grace.

An early version of chapter two appeared as "Composing for Sound: Sonic Rhetoric as Resonance" in *Computers and Composition*, vol. 43, 2017.

1 Resonant Listening as Cultural Practice

In July 2022, audio engineer and artist Jeff Rice recorded the sounds of an afternoon thunderstorm. What made the recording extraordinary was where Rice placed the microphone: next to the shared root systems of the planet's largest living organism, the 100-acre Pando aspen forest in Utah. Conservationists hoped Rice's recordings would give the organism a voice and, thus, a stronger claim to its land and resources. Listeners can access the recording, entitled "Beneath the Tree," on the web and experience the layered vibrational intensities of both the storm and the organism's response through Rice's carefully curated sonic composition. The artist and his funders want the listener to resonate and vibrate with the ever-shifting highly animated soundscape and then act in a way that supports and sustains this relationship, such as donating to the organization or voting to preserve it. Many recent sound art climate installations make similar rhetorical moves. That is, they amplify the resonances among humans and nonhuman elements in order to slow or halt irrecoverable damage to the planet's ecosystems. Mainstream audiences, they claim, have grown so desensitized to the red flag warnings of CO^2 charts and global temperature maps that only a multisensory experience via sound and multiple sensory modes will awaken people to their impact on the environment.

Such arguments, in turn, presume a particular form of listening, one informed by cultural and physiological normative practices. In his groundbreaking book on settler colonial and Indigenous ways of listening, *Hungry Listening*, Stó:lō Indigenous Arts scholar Dylan Robinson cites a 1987 legal proceeding where Chief Mary Johnson was asked to "sing a Gitksan song as an essential part of her evidence on the 'Avook,' the ancient but still effective Gitksan law." According to the report, Judge McEachern objected, saying "I can't hear your Indian song, Mrs. Johnson, I've got a tin ear" (37). Robinson frames this proceeding as just one example "of the many ways in which listening is guided by positionality as an intersection of perceptual habit, ability, and bias" (37). While the "tin ear" metaphor carries problematic cultural biases against certain forms of deafness, particularly tone deafness, Robinson explicitly employs the "tin ear" to distinguish between Indigenous and settler ways of listening in order to "trace the unmarked normativity of listening" and the ways certain forms of listening have been prioritized historically and culturally. One widespread settler form of listening, which Robinson emphasizes throughout the book, is the ability "to capture the content of what is spoken" or to capture information "over the affective feel, timbre, touch, and texture of sound." Robinson argues that "[a]ttending to affect alongside normative listening habits and biases allows us to imagine (or audiate) otherwise—to develop strategies for different transformative politics of listening that are resurgent in their exploration of Indigenous epistemologies, foundations, languages, and sensory logics; or, ones that are decolonial in their ability to move us beyond settler listening fixations" (38). On a broader level, Rice's recording does enact that hungry settler's colonial desire to capture, extricate, archive, and, thus, objectify the sounds (and organisms) of "nature." Objectifications of "nature" and "culture" ring false when examined through Indigenous and materialist feminist ethical stances that collapse false binaries between nature and culture (Alaimo 9). On another level, however, Rice's recording resists western cultural logics of syntax or melody, asking us to resonate with the timbre and texture of the organism's vibrations and dissolve settler fixations on linguistic content and western dominant musicality.

Much like Rice and Robinson's work, this book, *Composing Resonance: Culture and Collaboration in Sonic Practices*, approaches sound and resonance as cultural and rhetorical phenomena, which can both subtly and even profoundly shape our sense of self, other, environment, and community. We argue that more cultural and rhetorical research needs to focus on sonic practices within diverse locations, environments, and communities and illuminate how those practices challenge the "tin ears" of cultural listening biases that reinforce conventional notions of identity, including basic divisions between the self and others. This book aims to address that need for diverse locations by documenting the ways, through sonic practices, that particular sound artists, singer-songwriters, and students understand and compose their relationships to their bodies, environments, and communities. We also examine the ways these sonic composers participate within their communities and cultures in ways that reinforce and challenge normative ways of listening and producing sound. Many of the sonic pedagogies and practices described and analyzed in this book interrupt western habitual ways of listening to, capturing, isolating, and amplifying certain sounds over others; by doing so, they invite a wider repertoire of sounds and listening bodies into the soundscape.

Sonic rhetoric has emerged as a cross-disciplinary field over the past twenty-five years and has focused on sound studies within the contexts of sonic compositions, pedagogical practices, rhetorical effects of sound, and the phenomenology and materiality of sound. This book builds on that work and analyzes explicitly the cultural aspects of sonic phenomena as they are manifested in specific locations and local ecologies. We do so through a series of case studies, each focused on a location and cultural community that coalesces around and through sound. Each located experience exemplifies the receiving, amplifying, recording, and exchanging processes that co-produce resonance, and those sonic experiences demonstrate the essential links to culture's role throughout the listening and composing processes. The chapters move from the more familiar episodic, routinized spaces of the classroom to the ritualistic, regular rhythms of the songwriters' events, before examining the one-time sonic art event Re/Call. The movement from regular sonic practices to event-based ones highlights how sound shifts our sense of temporality, as well as our sense of space and embodiment. The final chapter then reckons with the limitations of the two main case studies and offers extensions for new activist research.

The purpose of this first chapter is to explicate the theoretical and disciplinary contexts for our case studies, detailing the key concepts of resonance, collaboration, location, and listening via a recursive interactive process. We show how this process is reimagined and circulated through culture. Like the "tin ear" of the settlers toward an Indigenous song, our inability or unwillingness to perceive cultural differences must be corrected and attuned to the resonances of a new sonic experience through the processes of receiving, amplifying, recording, and exchange—processes we observed in our case studies. The subsequent chapters exemplify that aim and examine the above concepts and processes while bringing a new conclusion or take away from each case study.

In chapter two, we discuss the sonic heuristics we undertake together with our students to interrupt and interrogate dominant listening practices. The chapter will be of particular interest to composition teachers and scholars, as it updates the sonic pedagogy we introduced in our 2017 *Computers and Composition* article "Composing for Sound: Sonic Rhetoric as Resonance." This version integrates resonant listening practices with a carefully sequenced set of heuristics and assignments, which move students from resonating with sounds as objects to resonating with themselves as listening subjects to finally resonating with a larger cultural sonic field. As a result, the sound object, the vibrating subject, and analysis of the larger cultural soundscape play critical interdependent roles in the students' final sonic productions.

Chapters three and four locate sonic practices within collaborative professional sonic rituals and events. In chapter three, we investigate the collaborative sonic practices that Mary and a group of singer-songwriter colleagues used to build community and challenge the traditional cultural notions of songwriting as rooted in individual genius, divine intervention, or production driven by future consumption. Appealing to teachers, scholars, artists, and musicians, the songwriters' study analyzes the collaborative exchange practices that singer-songwriters develop together through ritualized song circles and recording studio sessions where music producers mix songs within crafted hybrid (physical and digital) studio spaces. Like Robinson, the songwriters recognize the operations of dominant cultural assumptions and colonial fixations on capturing, producing, and preserving music for posterity. In their sonic practices, the songwriters' sessions work against those norms to promote ulterior listening norms for their specific subculture of Americana folk musicians as well as for this small collective group. Our analyses in chapter three apply James Britton's foundational concepts of "exploratory composing" and transactional improvisation to frame how songwriters' embodied experiences and collaborative practices characterize their songwriting. Mary and her musical colleagues engaged in resonant listening and ritualized cultural practices of receiving, amplifying, recording, and exchanging their songs to form their central sonic experiences. Those practices, in turn, offer us new lessons about pedagogy for self-expression, collaboration, and audience response.

In chapter four, we discuss how the creation of a sonic land archive in Michelle's Re/Call workshop brought cultural assumptions and tacit sonic practices and soundscapes into critical relief. Re/Call was a three-day sound art retreat/exhibit at the Rocky Mountain Land Library in South Park, Colorado, where Michelle was invited to present a soundscape workshop and curate participants' sound files into a digital archive. Drawing upon the experiences of teachers, scholars, and artists in the workshop and at the exhibit, chapter four demonstrates how participants used listening practices to collectively locate themselves and others in the exhibit and on the land. The chapter engages and problematizes cultural theorist Henri Lefebvre's *Rhythmanalysis* to amplify the sound art practices at Re/Call. There was a desire within the workshop to isolate and record the rhythms of the ranch and exhibit with high fidelity in order to create a sonic souvenir of the event, thus enacting what Robinson calls "hungry" white colonial settler ways of listening that extract and contain linguistic information for future consumption (2). Despite enacting this cultural drive, participants did initially engage in the messy and entangled bodily practices of receiving, amplifying, recording, and exchanging in order to resonate rhetorically with the rhythms of the land.

Finally, in chapter five, we reckon with the cultural silences and white settler forms of listening in our research methods and offer extensions for several new projects, such as a local urban renewal podcast and the Displaced Aurarian campus history project, both of which address cross-cultural issues of inequity and access in our research. The embodied process and analytical category of culture offers sonic researchers a way to identify sedimented sonic and rhetorical habits that (often tacitly) colonize our bodies, habitats, and communities. These newer projects grapple with colonizing ways of listening within a more explicit social justice framework and offer models for ongoing research. After we describe research that can help answer the crucial quality of life and social justice questions raised in recent scholarship, we encourage other researchers to make social and linguistic justice central to sonic rhetoric in classrooms more generally. We hope our readers (sound studies scholars, rhetoric and composition scholars, teachers, students, artists, musicians, community organizers, and more) find resources in each of the chapters to create their own diverse set of sonic research projects and assignments. In the remaining pages of this chapter, we trace the concept of resonance through recent scholarship in sonic rhetoric and sound studies in order to develop a framework that situates sound as a cultural, environmental, and collaborative phenomenon and process.

Resonance in Sonic Rhetoric and Sound Studies

Rhetoric and composition scholars have historically recognized the rhetorical and aesthetic power of certain sounds, particularly vocalization, to affect writing and speech processes. For example, the widespread influence of Cynthia Selfe's "Aurality and Multimodal Composing" and Peter Elbow's *Vernacular Eloquence* demonstrates the ongoing importance of oral/aural discourse to writing studies. Both pieces have made our own research possible, as Selfe and Elbow have effectively challenged the dominance of print as both a framework and mode for composing. Recent work on sonic rhetoric and composing, however, places more emphasis on the extra-linguistic, nonrational, and ecological aspects of sound (e.g., vibrational qualities, tone, pitch, frequency, timbre, resonance, dissonance, distortion, noise, and circulation) in relation to other modes, senses, networks, and environments (see, for example, *Currents* special issue on "Writing with Sound," *Computers and Composition* special issue on "Sound in/as Compositional Space," and the *Harlot* special issue on "Sonic Rhetorics"). Jonathan Alexander, for instance, analyzes the works of Glenn Gould, the Canadian musician and experimental multimedia artist, and concludes: "Attention to such work might help us consider what can be done with sound and voice in the production of multimedia texts where sound and voice act beyond the textual—not just as metaphors for textual meaning making, but as materialities with their own particular rhetorical and affective affordances" (75). Steph

Ceraso's *Sounding Composition* and Byron Hawk's *Resounding the Rhetorical* push Alexander's argument further by problematizing the notion of sound as just another isolated mode of multimodal composition or as another tool in the rhetorical toolbox; instead, they both claim that sound, which is inextricable from its environment and technologies, offers us new ecological and collaborative frameworks for thinking about composition and rhetorical practices in general.

Ceraso constructs an ecological framework by challenging our reductive "ear-centric" approaches to composing with sound and introducing the rhetorical practice of "multimodal listening, or attending to the ecological relationship among sound, bodies, environments, and materials" (3). Such listening, she argues, moves away from simply hearing and interpreting "audible sound" and calls for "a bodily retraining that can help listeners learn to become more open to the connections among sensory modes, environments, and materials" (5). Sound, according to Ceraso, is "the ideal mode for exploring the multisensory nature of multimodal interaction" (9). Because sound is composed of vibration, it can be seen, felt, and heard, making it "an especially effective—and affective—mode for understanding how the senses work together *ecologically* in multimodal experience" (10). Like this book, *Sounding Composition* garners multisensory listening experiences and practices from an "eclectic mix of contemporary sound practitioners" (3), including deaf multi-percussionist Dame Evelyn Glennie, acoustic designers, and automotive acoustic engineers, who offer "everyday sonic spaces and experiences outside of what is typically thought to be the domain of multimodal composition" (143). At the end of each chapter, inter-chapters called *Reverberations* offer ways to extend these conversations into the classroom. By doing so, Ceraso demonstrates how "the bodily practices and habits that students develop via multimodal listening are relevant to a range of personal, professional, cultural, and civic activities" (145).

Hawk's *Resounding the Rhetorical* marks a similar shift from studying sound as a compositional mode or object to situating sound within a network of complex material rhetorical practices. Referring to his own early (2006) advocacy of sound as an important additional meaning-making mode, Hawk argues it is now time for a more "substantial rethinking" of sound as not attached to a particular mode or medium but as a "function of larger processes, ecologies, and mediations" (6). Hawk goes even further to suggest that sound, with its explicitly vibrational and emergent qualities, pushes the field to reconsider its very objects of study (e.g., rhetorical/ composition texts, websites, platforms, songs, and artworks) "as a series of processes not tied to any medium in particular—whether textual or visual, print or digital." The field's "object" of study or "quasi-object" (partially moving and emergent), a critical concept Hawk borrows from Michel Serres, then becomes "anything in the process of being composed" (6). Quasi-objects are, according to Hawk, "bound up with the energies that drive them, the ecologies that co-produce them, and relations that sustain and transform them" (7). In *Resounding*, Hawk's quasi-object is musical composition, which he defines as "the broader material practices of gathering of technologies, spaces, and relations to co-produce sound waves that are translated into a recorded format such as an MP3 file, performed live, promoted via social networks, and then distributed across the Internet, ultimately to be recomposed along the way into sound waves that gather a whole host of technologies, people, genres, and practices in every co-productive moment of reception" (8). Musical compositions thus live and morph with recompositions and engagements in rhetorical circulation.

Hawk also relies on case studies to examine the movements of musical composition: compositional practices in a recording studio, the human gestures of studio and live performances, the use of social media to compose publics, and the online circulation that transformed an underground album into a classic. His descriptions are ultimately "entangled with" an abstract model of musical relations ("recording, performance, promotion, and distribution" [10]) and key terms for rhetoric and composition ("composition, process, research, collaboration, publics, and rhetoric" [12]). He then reassesses these key terms through the nuances of each case.

Hawk's definition of resonance is particularly important to our work here. He traces it through its acoustic meanings, as "the capacity of a material structure, like a wall, to vibrate at a certain frequency" (33); its scientific meanings, as "the transfer of energy" and "gravitational influence" (34); and its humanistic meanings, as a phenomenological "model of being-in-the-world" (35). Ultimately, he uses the term to describe the workings of rhetoric itself. Drawing on Laurie Gries' *Still Life with Rhetoric*, Hawk argues that "[t]he rhetorical . . . resonates" (193). In her research on the Obama Hope image and its circulation through multiple emergent networks of association, Gries demonstrates how the rhetorical is more than simply a singular site of production or delivery. Instead, as Hawk explains, it "resonates, coproducing unforeseeable, divergent consequences in excess of intent that unfold through a distributed process of circulation" (193). "Through circulation," he continues, "resonance entangles conditions produced by past works, embodiment in present listening practices, and affective impacts on other resonant bodies and ecologies, even in distant futures" (193). Later, after tracing the affects and effects of resonance in each case study, Hawk situates material resonance ("the capacity of a body or wall to more easily absorb a particular frequency and vibrate") as co-productive of the rhetorical, as creating "a sustained ringing or repetition of the original tone or frequency to produce a deeper, fuller, more prolonged sound" into the future (224).

Our book aligns with Hawk's aims and his foundational premise—that "[t]he one constant across the cases that make up *Resounding the Rhetorical* is the ontological presence of sound, resonance, ringing as the ground of composition" (233). Our case studies are also grounded in this ontological premise with fully embodied composing practices occurring in additional professional and natural spaces. Resonance, as an inherently embodied and collaborative process, contributes to our sense of self, identity, location, orientation, and relationships. It is a biological and social phenomenon that crosses our conceptual self/other, inside/outside boundaries, creates our capacity to influence and be influenced by our environment, and undergirds our ability to make meaning. As Hawk states, "[Resonance] entangles materiality *and* meaning . . . [and] induces shifts in how humans live and dwell in the world" (193; our emphasis).

These shifts go beyond our learned experiences and material entanglements, as they operate as larger cultural forces in our environments. In a telling example, after observing the planet's seismic hush during the COVID-19 pandemic and subsequent lockdowns in 2020, Clemens Finkelstein, a PhD candidate at Princeton University's School of Architecture, used a small, cheap device called the raspberry shake, which measures "minuscule ground movements," to make seismology more accessible to the public (Morley). The raspberry shake that Finkelstein placed inside the CIVA (Centre for Information, Documentation and Exhibitions) in Brussels could "show people coming into the space the vibrational impact they have on their surroundings," and some visitors took the liberty to stamp their feet to actively engage with the sensor. The show "unpack[ed] the multiple ways in which ideas about sickness inform architecture" (Ayers). As Finkelsten explains:

> [C]oncerns about human and planetary health have always been closely entangled, with architecture as a mediating agent between cosmic and biological bodies. The standstill of human activity that occurred when government lockdown protocols turned cities into ghost towns during the COVID-19 pandemic also slowed down the ubiquitous vibrations of the Earth by up to fifty percent and culminated in the longest period of seismic quiet in recorded history. If Bezos, Musk, and Co. have their way, one day we'll leave behind our sick world to infect the rest of the universe . . . (Ayers)

Citing historical examples of sanatoriums as designed for sick bodies, the "Sick Architecture" exhibition, which housed Finkelstein's installation, demonstrated how sonic vibrations, "architecture and sickness are tightly intertwined" (CIVA). "Architectural discourse," according to the exhibition's brochure, "weaves itself through theories of body and brain, constructing the architect as a kind of doctor and the client as patient" (CIVA). Understanding built space and seismic vibration as a health indicator—for the planet and all organisms—situates the modality of sound as central to experiencing and responding to climate change and environmental devastation. Sonic vibrations' centrality to such experiences also encourages scholars in various disciplines to dissolve their fixation on sound as merely verbal linguistic information and content. Like Ceraso's examples of bodily retraining, listeners attuned to the meanings conveyed by architecture can then experience new connections among senses and material environments.

Where this book departs explicitly from Hawk and Ceraso is in its emphasis on critical environmental listening alongside a proliferation of the subcultural practices and ideologies that inform how we perceive and attribute meaning to sound. For example, it is through cultural training that we notice, differentiate, and project a particular sound based on assumptions about the "voice" of a cisgender white woman or the "voice" of a transgender person of color. Through this localization and identification process we, the listeners, also assume a particular shape and sense of our own self. If we perceive a sound as threatening and directional, our sense of individual self tends to become stronger and more confined to the so-called physical body and may move us into instinctual fight or flight mode. Conversely, if we perceive a sound as beautiful and immersive and safe, our sense of self becomes more expansive and less defined. Whether we are habituated to hear a sound as threatening or beautiful is, in turn, dependent upon cultural experiences and deliberate training. As musicologist Nina Sun Eidsheim argues, the ways we listen are cultural and mirror the ways we interrelate with the world in general (2). These ways of listening are made habitual and seemingly natural through repeated sonic experiences to the point where we notice certain sounds over others (11). Habituated sonic practices are what make sound a particularly powerful tool for noticing our cultural assumptions and norms around identity and community. In this sense, our work also builds on recent research by Cecilia Valenzuela and Magnolia Landa-Posas. Their panel presentation published in *Kairos*, "AudioVoice: A Relational, Subaltern Praxis of Listening to Testimonies and Composing with Sound," gives voice (*testimonio*) to (im)migrant multilingual communities living at the borders of gentrification. By composing soundwalks throughout gentrifying neighborhoods in the Denver metro area, these authors juxtapose family stories and sounds with the ambient sounds of "development" and "progress." These layered sounds expose shifts in cultural and racial identity and values as new housing, retail, and transportation centers are built and pollute the neighborhood's soundscapes.

As Casie Cobos et al. discuss in "Interfacing Cultural Rhetorics," the "concept of culture is taken up in a variety of ways in rhetoric and composition scholarship" (141). Rhetorical inquiry, they argue, tends to define culture as an object, as context, as a shared process, as collective ways of making meaning, as values or beliefs, or some combination of the above, and it tends to "locate" culture in languages, customs, identities, and/or forms of organization. This concept of "culture" has been rightly critiqued in anthropology and sociology for its valorization of authenticity and essentialism (Clifford and Kahn). For example, Malea Powell et al. discuss how this reductive use of the concept has affected the field of rhetoric and composition: "All too often, scholars in rhet/comp rely on this object-oriented approach to cultures because it allows us to select 'exemplars' from specific oppressed cultural traditions as a way of feeling good about how inclusive our discipline has become." Like Powell et al., we invoke "culture" to resist such exemplar objects and instead discuss the everyday messy processes by which *we all* (researcher and researched) acquire habits, beliefs, and values and build communities. Culture is both an embodied process and an analytical category that helps us identify those sonic practices, values, and beliefs that are shared among a group and become "naturalized" over time through repetition and enculturation. For us, culture is not an object out there, with strict place-based boundaries, nor is it located only in formal rituals, recipes, and language. It is, instead, a naturalization process—what LeFebvre called "dressage"—experienced in the everyday, a concept we explore further in chapter four (16).

Enculturation—whether it happens at a local or global scale—is never complete and changes with each individual's everyday encounters with material conditions. In fact, this inherent instability makes it possible to identify and transform those material conditions (Powell et al.). We agree with Eidsheim's argument that it is important for scholars researching sonic composition and production to deconstruct sonic practices and beliefs in order to understand how we value certain sounds over others and how that process, in turn, signals our relationship to other sensory experiences (seeing, tasting, touching, and so on). Ceraso's focus on whole body forms of listening vs. ear-centric forms of listening explicitly implies cultural distinctions between the two. Thus, by employing culture as a primary category of analysis, we can explicitly deconstruct these forms of listening processes. Like Ceraso's "sonic scenes from everyday life" (2), where she describes in phenomenological detail her experiences of listening in a coffee shop, on campus, and in her car, our own descriptions of embodied listening as sonic scenes are an important first step in our research. To understand in minute detail the ways we process and experience sound is to understand the ways we interact with the world more generally—what Eidsheim calls a choreography of embodied actions.

Our next move for our case studies research, though, is to ask ourselves, our students, and our co-participants how these listening practices are informed by cultural, racialized, and/or gendered norms—in other words, how these experiences reproduce the entrenched notions of which sounds are deemed important. Just as our ways of seeing do not emerge from a blank slate, neither do our ways of listening emerge from silence. In fact, as Eidsheim stresses, our bodies are makers of sound through amplification, transmission, and transduction. To understand ourselves as embodied subjects that are affected by various sonic objects does not account for the equally important ways our bodies co-produce sound in dynamic relation with others and build our experience of resonance within specific locations.

Drawing attention to the cultural elements of sonic experience, including the assumption that listening is a passive activity where we simply receive and process sounds within our bodies, makes sonic practices available for critical reflection and transformation. Such attention exposes how our individual sensory processing and meaning making are part of larger cultural processes and rhythms, instead of reinforcing the inside/outside binaries of a westernized approach to the body and its environment, an approach that, in turn, reinforces the very notion of an individual body or voice emerging from the silence or noise of the proverbial wilderness. Furthermore, it allows us to hear the sounds and voices marginalized by those in power, who continue to silence and exploit whole populations of humans, animals, and ecosystems for profit. This exploitation is hardly new, especially given the climate and pandemic disasters multiplying in the twenty-first century, as well as sociocultural inequities within the rise of white nationalism in politics, media, the military, and police organizations around the country. Even more pronounced is a growing skepticism—even among those outside activist circles—toward individual-based thinking and action in the face of large-scale exploitation. We add to this skepticism by questioning the very ways we use sound and hearing to differentiate self from other and human from environment. Unexamined dominant cultural listening reinforces individual and group control, as well as domination of the environment and others.

Collaboration and Location-Based Composition

Valenzuela and Landa-Posas identified and challenged cultural norms through a subaltern perspective, while Eidsheim challenged them through experimental music and art. We challenge cultural norms through critical composition pedagogies and collaborative art- and music-making practices. This work is feminist and environmentalist in that it exposes and challenges colonial, gendered, and racialized ways of listening and knowing. Our collaborative participant-researcher projects occurred in conjunction with an outdoor environmental art exhibit and workshop conducted in Colorado and with a mostly Atlanta-based group of songwriting musicians who produce and perform original music. The participants in the environmental art exhibit Re/Call gathered at an evolving land archive called The Rocky Mountain Land Library at Buffalo Peaks Ranch, an historic ranch near Buena Vista on the banks of the South Platte River, and the collaboration of singer-songwriters occurred at two small professional music studios in Atlanta and Asheville, NC, as well as group song circle events. These case studies are based in highly collaborative processes: Michelle met on the land over the course of several seasons with curators and the participants to plan and co-produce the workshop, while Mary met with other songwriters for song circle gatherings and in the music studio to co-produce recordings. Our location-based case studies identified sonic practices that encouraged participants to resonate and empathize with each other, as well as the broader rhythms of the land and local cultures, in order to develop a sense of place, community, and living history. Thus, while we describe sonic experience in a phenomenological way much like Ceraso's recordings of the physical and emotional impact of particular sounds on her body, we frame these experiences as forms of participation, as forms of community and world-building.

Because we focus on the local and cultural conditions for consuming and producing sounds, we also draw on research on sonic cultures and practices. A growing number of scholars working in sonic rhetoric, including emerging scholars, are approaching music and sound art from a community and cultural perspective—a perspective that offers nuanced views of local people and practices. Jonathan Stone's work is particularly enlightening in his focus on vernacular folk recordings and their significance for rhetorical studies. In his study of John and Alan Lomax's 1933 recordings of proto-blues songs sung by African Americans in eleven Southern prisons, Stone highlights the ways the recordings—with their dissonant and multivocal qualities—challenge the conventional notions of African American identity among their white middle-class audiences. He writes:

> Institutionalized or nationalistic symbolism codifies African-American experience (and race itself) into a reduced and simplistic monotone. . . . [T]he institutionalization of the prison recordings has the same effect as their digitization: compression, distortion, and the codification of various imperfections. On the other hand, the sonic symbols at play within the vernacular context of the song itself—represented nicely in the antiphony of call-and-response—is that of community, sympathy, and shared struggle. The former symbolization is reductive, the latter productive.

We also agree with Jordan Frith's argument that location-based composition is a multisensory layered phenomenon, with multiple simultaneous locations in physical, digital, and textual spaces. That was, indeed, the case in one of the Re/Call installations, which utilized a form of locative media—a smart phone sound app—to direct drivers and passengers through the southwest Denver metro landscape toward the physical site of the ranch. Frith writes, "Location-based composition . . . does not divorce the description from the thing being described. The textual description . . . becomes a part of the thing being described and opens up new rhetorical potential for mobile composition" (52). This "merging of digital information and physical space" then influences "the ways people 'read' and 'write' physical space" in general (52). Nathaniel Rivers makes a similar argument in "Geocomposition in Public Rhetoric and Writing Pedagogy" but further explores the ethical and political implications of locative media and writing—in particular the practice of geo-caching. Geo-caching, a collaborative form of movement through a public space, according to Rivers, is a form of rhetoric, a form that also determines who gets to go where and why (585). For the songwriters, the layering of shared spaces moves among several locations where singing voices, played instruments, recordings and performances comprise the music jam or song circle, the recording studio, and the smaller music venues where independent musicians tend to perform. Each location intermingles with other events in a layered dynamic such that each part of the process has a sonic rhetoric that moves participants into more ethical and collaborative interactions. Such values were also at work in the chapter four case study, where not only the smart phone app, but also the river chimes and sound mirror installations, became powerful forms of sonic rhetoric, moving human visitors into more ethical relationships with the land and its nonhuman inhabitants.

When we apply the terms "location-based" and "community" throughout this book, we are not referring to landmarks on a map or static points in a room; instead, we are interested in how sound, as a rhetorical phenomenon, shapes experiences on a continuum of location and dislocation, of community and alienation, and of privilege and marginalization. In some contexts, like the Black Lives Matter protests of summer 2020, sound (e.g., the sounds of rubber bullets being fired, tear gas canisters hitting the ground, the heavy drones of police riot vehicles and helicopters) is deliberately used to dislocate, disperse, or even destroy community groups. Likewise, sound can influence experiences of communion and community through shared sonic rituals (e.g., music festivals, sing-alongs, and even social media dance challenges) within particular physical and digital locations. Thus, we are also interested in the processes we use to locate sounds or contain and map sonic events. Sound researchers call this process "localization," which involves complex habitual ways of interpreting sonic distance and direction (Blauert 19). As a cultural, as well as physiological, process, localization informs the ways we navigate, imagine, value, and compose our everyday soundscapes.

Resonant Listening Practices

Whether we are navigating a natural landscape, recording stories for an archive, or exchanging songs in a song circle, fully embodied "multimodal listening" (Ceraso 3) —listening through the skin, ears, bones, and/or auditory processing networks in the brain and body—is a primary way we orient and locate ourselves within our communities and environments and respond to others. Along with its orienting capacities, sound also has the power to dislodge or interrupt our usual meaning-making processes. In fact, it is often through our failure (or refusal) to identify, control, and "localize" the source of a sound, that we can see our own culturally habituated attempts to make meaning and begin to question those processes. Physiologically, the very experience of hearing sound emerges from forces largely beyond our conscious control: air molecules and waves, ear drums, haptic nerves and brain processing, and social regulations/conditioning and training. Simply moving from one part of the room to another or even adjusting our posture automatically (and often, unconsciously) changes the way we process sound, with movement itself dampening our own auditory responses (Livneh and Zador 18). We thus emphasize "resonant listening" as a rhetorical practice and sound as a rhetorical resource, though we do so knowing both rely upon complex, collective multisensory processes beyond our control.

Resonant listening recognizes the vibrational qualities of all objects and bodies and how one vibration easily influences, amplifies, and transforms another. Such recognition helps one 1) notice the amplitudes and frequencies, the rising and falling qualities of particular sound waves as they are emitted (what we call "reduced listening," a term and phenomenological process borrowed from Michel Chion); 2) notice how our bodies, environments, and instruments receive and amplify particular sonic vibrations in particular ways (what we call "embodied listening"); and 3) notice how we record, remix, and exchange these vibrations within the larger cultural rhythms of the surrounding landscapes.

We recognize that our human experience of vibration, sound, and resonance is stunningly diverse and dependent upon cultural, as well as biological, conditioning. For example, certain musical chords may sound like dissonance or noise to a person trained in western classical music, while they sound harmonious to a person trained in eastern classical music. We also engage in diverse bodily awareness practices—practices that choreograph the ways we process sound in and through our bodies—which depend upon dominant ideologies about the body (e.g., body as instrument, body as computer, body as self-organizing system, body as transparent vibrational field, etc.). In his study of Enlightenment notions of the body as instrument, music historian Samuel Breene outlines the effects this concept had on listening and composing practices at the time. One of the more popular Enlightenment accounts of Mozart's "embodied state" while composing described it as "a set of strings, harmoniously arranged with such art that a single one cannot be touched without all the others being set in motion" (234). Musicians and audiences of the time, therefore, sought to develop similar embodied sensitivities and gestures tuned to the harmonizing resonances of piano and violin strings. Like Enlightenment Europe, many western eurocentric cultures continue to value the ability to discern music from noise or even ambient sound from foreground music. As Eidsheim argues, the pressure in these cultures is to ignore our multisensory experiences of sound and turn isolated sounds into stable mono-dimensional referents for something we call "noise" or "music" (2). To illuminate and thus expose this reductive cultural tendency, we emphasize resonant listening practices that are multisensory, relational, and located in specific, yet dynamic, environments, places, and bodies.

Within this diversity of locations, general meaning-making activities related to sound emerged for us and the other participants. These embodied processes included:

1. A sense of location via echo and reflection of sound waves within a space
2. A sense of time passing with the rhythmic cyclical rise and fall of sound waves

3. A sense of movement/stillness with the rhythms (expansion and contraction) of sound waves
4. A sense of self/other through the identification of distinct voices
5. A sense of well-being/disturbance through the recognition of sonic density and diversity vs. low frequency monotones or silences

All these meaning-making processes are filtered through our senses and cultural conditioning and do not necessarily or automatically lead one toward an ethic of caring for the environment or for others, but they do entangle listeners with the land, the architecture, and its inhabitants. This book focuses on how artists, musicians, and students have collaboratively developed practices in resonant listening and worked together to shape (albeit imperfectly) powerful semiotic phenomena.

Resonant listening, thus, informs the collaborative practices of sonic composing —receiving, amplifying, recording, and exchanging—that operated both in Michelle's environmental art exhibit and in the songwriters' recording sessions and gatherings. Those same practices then became part of our composition classrooms, where students used them to develop effective multimodal and digital writing projects. Emphasizing such practices can help students 1) identify the individual and cultural ways we use sound to locate and orient ourselves and 2) record and remix sonic elements for the purpose of moving audiences toward a more diverse, inter-relational sense of place, community, history, and the environment.

Receiving

The practices of receiving we noticed in our two case studies included Ceraso's multimodal, embodied listening practice. We labeled this practice "receiving" in order to highlight the body's role as a literal receiver/producer (the two are inextricable) of sound, much like an actual radio receiver that uses an antenna to capture radio waves in the atmosphere, then extracts those waves that vibrate at a certain frequency. Such a practice requires a set of inner gestures in order to create a capacity to sense through the ears, bones, and cavities within the body a diversity of subtle, as well as intense, sound vibrations or sensations. The attention is placed on the sensations or physical feelings, as opposed to the perceptions or mental images, that accompany contact with sound waves. Before even identifying or localizing the sounds, one can attend to the vibrations and their impacts on the body, including whether parts of the body expand or contract in relation to those sounds. Ceraso, for example, asks her students to design a digital media project or performance that conveys such an experience. Students were to convey the feelings and body sensations that accompanied an experience of sound—whether it was a plane taking off or a local hip-hop concert. Her assignment cultivated the receptive awareness we noticed in our case studies. In Michelle's case study, such receptivity was not always a pleasant experience nor was it devoid of cultural conditioning, as participants registered the impact of traffic, wind, and the cries of ranch cattle on their physical sensations, feelings, and thoughts, but it was a necessary step for mapping the soundscape of a local environment.

One could argue that having a body presupposes the ability to receive and be influenced by the environment. In fact, receptivity is essential to any rhetorical practice, especially if we define "persuadability" as rhetorician Karlyn Campbell does: as the "capability of influencing and being influenced" (103). Each encounter with a sound, as it vibrates our eardrums, bones, cartilage, and neurons could be understood, using this definition, as rhetorical and as influencing our sense of self, other, and environment. If a sound, like an oil tanker truck passing at eighty miles an hour on a two-lane highway, causes the body to contract, that gives the listener a strong sense of identity separate from the noise and by conjecture the truck itself. If, on the other hand, a sound like a river flowing over rocks causes the body to expand, that gives the listener a more fluid sense of identity, perhaps as a participant body within a larger ecosystem. Again, while these were common experiences for the participants in Michelle's case study, they are also strongly influenced by social and cultural conditioning. It could be argued that the sound of a speeding oil tanker truck causes a particular human body—one that benefits from such commerce—to expand to a sense of a larger network of circulation and distribution of resources and social privilege.

The ability to receive a diversity of sounds, including the roar of semi-tractor trailer engines, and the ability to recognize the body as a receiver of persuasive sensations, are crucial to sonic practice and can be cultivated through a number of techniques, including reduced listening. We both used this training with participants in the case studies and with students in our sonic rhetoric and writing courses. We explain this pedagogical technique more fully in the second chapter of this book, but we want to highlight its importance to the practice of receiving here. Our approach to reduced listening, which is based on Chion's phenomenological heuristic, asks listeners to pay attention to the physical qualities of sounds before identifying their sources or meaning. Since this is largely impossible, especially in the beginning, we ask listeners to note at what point in the hearing process they move toward the work of naming the sound and interpreting its meaning. While sonic composers may cultivate receptivity to a diversity of sounds without some form of reduced listening training, it is not easy since identifying the sources of sounds (the localization process) is usually an integral part of our survival instinct and largely unconscious cultural conditioning and is often triggered by past traumatic experiences.

Similarly, the sonic techniques and everyday practices used by the group of singer/songwriters who write, record, and perform original music typically begin with moments of receiving and other processes of resonant listening. Our argument that sound is an indicator of culture and a cultural soundscape's meaning emerges in the process of songwriting and begins when songwriters first receive and resonate with certain sounds in themselves. Songwriters sometimes describe receiving musical compositions wholesale, as if they were merely channeling that music from "elsewhere," while others describe coming up with words, tones, melodies, beats, instrumental riffs and attention-catching hooks in bits and pieces that they can then weave together. Thus, in Mary's case study, receiving takes on explicit, as well as tacit, forms. In her own experiences with writing, performing, and recording music, groups of songwriters typically receive parts of the lyric and the musical arrangement, both tacitly and explicitly, via their instruments, voices, or software. They gather with other musicians who want to play together, but also to support one another's songwriting, to perform parts on one another's recordings, and sometimes to perform live together for their audiences.

Amplifying

The practice of amplifying might seem a bit more strategic than receiving, but it, too, has tacit dimensions. We may unconsciously resonate with, and thus amplify, certain sounds over others, due to our cultural and social conditioning. Some of this ability is physiological: the human body, due to its material makeup and cognitive processes, resonates with a relatively small (compared to other organisms) range of frequencies (Williams). For example, the body itself might resonate with and tacitly amplify a low frequency bass tone through the chest cavity, where the vibration becomes a pounding pulsation. Again, this capacity to amplify is dependent on cultural training, as well as anatomical factors.

Our use of the term "amplify" includes these tacit dimensions but also emphasizes the strategic amplification of particular sounds (and voices) over others. Unlike the other processes, amplification has roots in classical rhetoric. *Auxesis*, the Greek term for amplification, emphasizes intensity—the audience's attitude toward the subject—more than the use of elaboration or expansion. In his *Rhetoric*, for example, Aristotle associates amplification not with elaboration but with speeches of praise and blame: "Since the actions [of the subject of praise or blame] are taken for granted, it only remains to invest them with magnitude and beauty" (1368a). Particular methods of amplification, according to Aristotle, include emphasizing the actions of the subject as singular, as greater than the actions of others in terms of virtue or number. In deliberative and forensic arguments, Aristotle advises using amplification to emphasize the importance of your argument after you have established the facts (1419b).

Auxesis thus emphasizes the affective and energetic aspects of amplification. It attends to the emotional energy an audience already brings to a subject or argument, while in sound engineering, amplifying refers to the energy a producer of sound brings to the act—beating a drum with more vigor, blowing harder on a flute. Adding electricity, microphones, and volume (hi-fi amplifier) gives the sound input a direction, makes the electric currents larger, and sends electrified sounds to a loudspeaker. The speaker then makes sounds into much larger vibrations. The hollow sound box of a guitar or violin also works to reverberate and redirect the amplified versions of the small sounds of the instrument's strings. The dynamics of musical arrangements to create contrast—as musical elements move between louder and softer volumes, for example—depend upon these kinds of amplifications. Even cupping a hand behind one's ear funnels and enlarges environmental sound. During the Re/Call sound workshop, participants deliberately cupped their ears toward and placed their smartphone microphones close to less accessible sounds—less accessible because of the overarching sound of the wind or because of their location away from the human activity of the exhibits. One participant created amplification by throwing together a makeshift hollow box out of a pile of wood, which allowed the wind to bang two of the pieces together within a larger hollow.

The workshop also emphasized the cultural, and largely unconscious, ways we amplify certain sound objects over others (*auxesis*). Certain sound objects come to us with already intense affective and physiological charges—what neuroscientists call salience—and the workshop participants discussed that process together. Auditory salience, which brings our attention and reaction to a certain signal, still vexes acoustical researchers. Why, for example, does a certain sound object—a train whistle, for example—reverberate in our chest or in our stomach, while a bird song might seem to reverberate behind our eyes? One workshop participant described the sounds of a recorded steam engine as comforting, as rhythmic with her breathing and digestion, while another felt it in his head, with a tension and desire to block the "noise." For the former participant the rhythms of the steam engine emerged as prominent, while the other noises—the whistle and the wheels—withdrew into the background. These amplifications and their attending withdrawal processes are partly due to the length of the sound wave but are also indicative of cultural and affective training. What sounds do we consciously experience and what sounds are automatically shut out or diminished? Thomas Rickert argues for the importance of including such processes—the processes by which salience emerges from ambience—as part of the rhetorical situation: "An ambient rhetoric continually attunes itself both to what is present and to what withdraws: they are the conditions that give rise to our ongoing perceptions and understandings of the world" (xiii). For Rickert, such attunement requires attending to the role of emotion and the material environment as integral to rhetorical theory. For us, this includes attending to our sensory and physiological experiences and histories, which help us track, however inadequately, the ways we perceive our environments and then orient ourselves within them.

Songwriters also tap into individual, cultural, and historical experiences as they decide which parts of the composition are salient for a process of amplification. While playing a song with other singers and instruments, what remains foregrounded in the musical composition, and what drops away? Combining voices with harmonies amplifies the vocal line. Furthermore, many songwriters compose collaboratively; and in a songwriters' group, each instrumentalist may offer a new musical element or a partial song with the explicit goal of receiving feedback on what to amplify or what to add for strengthening the song's impact on listeners. Before they can fully collaborate with others to record songs, however, songwriters must first amplify what they, the others, and also the environment have determined need emphasis.

Our perceptions and experiences affect what we amplify, and for many marginalized populations, particularly Black transgender and cisgender women and men, who have experienced violence at the hands of police, the sound of a police siren can signal life-threatening danger. As Jennifer Lynn Stoever argues, the sounds of police sirens, police dashboard camera videos, and bystander cell phone videos of police violence "have histories." "If we listen," she writes, "we can hear resonances with other times and places: segregation's hostile soundscapes, the obedient listening that whites expected of slaves, the screams and prayers of Frederick Douglass' Aunt Hester. Even with mainstream news coverage insisting that each event—though dispersed by geography and circumstance—is "not about race," the sonic color line inextricably connects them" (3). The sonic color line is just one powerful example of the ways historical experiences of trauma inform salience and our amplification practices.

Recording

Focusing on recording practices allows us to talk about a particular piece of sound technology—the microphone—whose primary job (depending on its structure and placement) is to capture discernable and recognizable sounds and rhythms over random noise. Of course, microphones are only part of the recording process, which involves inscribing and storing sounds for future use (e.g., reproduction, distribution, archiving), but they have been such an indelible part of the semiotic and rhetorical processes of sonic composition in our case studies and in our classrooms, we want to highlight their important functions for recording here. In his discussion of the importance of microphones for audio-visual texts, Bump Halbritter expresses both horror and delight over the microphone's fickle role in capturing sounds, recounting the ability of one particular microphone to pick up the sounds of Andrea Lunsford's necklace over the sounds of her voice (123). Initially arguing that microphones cannot pay attention, Halbritter changes his point slightly to say they can *pay* attention (depending on how they are made) but cannot *shift* attention (123). Filters and settings make it possible for most microphones to "pay attention" to a particular spoken voice, like Lunsford's, or the full sound of a cello over the "noise" of people walking by or chairs scraping. Frequency level also determines whether or how a microphone picks up a bass or a soprano voice. However, as Halbritter humorously admits, any microphone, despite its programming, is itself influenced and transformed by multiple "actors" within any environment.

Similarly, in *Resounding* Hawk details the number and placement of microphones during a studio session, arguing that microphones provide multiple points of conversion within a larger nonlinear transduction process where sound waves are turned into electrical currents, then back again and so on. According to Hawk, the technologies involved in this process "aren't static; they are active emergent co-participants in the production of sound that also change, evolve, intensify, heat up, or slow down as a part of the transduction process" (101). As Hawk emphasizes, all participants—technologies and human singers and listeners—are changed, and in rhetorical terms, *influenced* in the process, whether or not they are conscious of the many major and minor transductions and transformations.

During the Re/Call exhibit, recording was a difficult collaborative practice, where participants (without microphone muffs or contact mics) contended with their microphones' ability to transduce the sounds of the wind over all other sounds. (The wind-generated noise with its higher air pressures will override other sounds without a physical windscreen, and such noise is also hard to correct with post-production software.) The situation foiled their attempts to capture and collect distinct sounds and complex vibrational experiences (such as a sonic yoga sleep meditation) from the exhibit and ranch; instead, they were left in awe of the overwhelming sounds of the wind and an occasional bird tweeting or gate creaking in response to it. The collective listening experience encouraged them to take their "shoddy" microphones and record smaller shielded enclaves on the ranch property. Some of them wanted group members to identify the sounds while others hoped to elude any easy recognition. The group's goal—to individually contribute to a published sonic archive of the event—supported the brief recordings of burrowing by the river, walking through the hay outside the barn, and so on. Isolating these particular sounds from the environment was also a cultural process in that they enacted sanctioned and valued human interaction with the environment. The cultural and technological effort exerted in capturing any sound other than the wind became part of the recording process itself, and the strong influence of the wind (one among many effects of a rapidly changing climate) on their bodies, soundscapes, and technology eventually became the main point of the exercise.

A phone's voice memo application or a portable audio recorder functions for many songwriters as a composition tool with its basic omnidirectional microphones. Singer-songwriters often create what are called "scratch tracks," which are recordings with a simple voice part and accompanying instrument played at a steady pace (e.g., to a metronome or click track). Songwriters then use those informal recordings as a rhythmic outline that indexes a template for the song's composition. Their tools act as a commonplace book to capture an informal collection of fragments and ideas that can become the basis of a song. Studio settings, with their extensive recording equipment, audio editing software, advanced microphones, and sound isolation, might be places where a musician typically fine tunes the finished recording. In a professional music studio, the music producer and/or engineer sits at a desk in front of a computer running audio software, while the musician records in the soundproof recording booth with carefully placed microphones. Capturing sounds, instruments, and voices separately into isolated recorded tracks is the routine process that allows for complex nonlinear mixing of musical elements, yet each take is a dynamic encounter among the technologies, humans, and physical spaces to enact what Hawk describes as a dynamic transduction process for recording with different kinds of microphones. Re/mixing is the ongoing recursive process of hearing "the room" (e.g., the studio, the performance space, the participants, and their affective responses), selecting recorded elements, adding sonic effects, layering sounds, and finally, sequencing or blending them together into a "final" (for the moment) mix.

Recording in the studio ultimately becomes a collaborative process of adding instrument parts and other sonic elements and creating effects through software (e.g., reverb or echo to soften the sound). Working for years with singer-songwriters who have recorded and published their original music taught Mary how the process in a music studio, taken as a whole, most resembles a professional publications project, where production and distribution of the song is the primary original goal for songwriters. However, this group of songwriters used these processes and went against expectations, explorating and collaborating along the way so as to de-center the fixed goal of a "final" song.

Exchanging

Exchange is defined as a transaction, usually by agreement among two or more people, whereby each party receives something similar in type or value to what they give. In this book, exchange is a metaphor for the deeply collaborative practices of sonic composition, where each participant is engaged in ongoing processes of exchange without a definitive beginning or end, and they do so in a specific highly ritualized and choreographed event when participants (human participants but also the architectures and technologies of a space) give and receive bits of recording, musical notes and phrases, layers and tracks, words, and meaning. Our use of the term exchange relies on studies in rhetorical circulation by Gries and Collin Brooke (*Circulation, Writing, and Rhetoric*) and Hawk, whose research examines non-linguistic conditions for rhetoric and the role of the nonhuman in sonic rhetorical practices within a vital ecology of circulation. "Circulation," according to Gries, "has been especially crucial in developing new rhetorical models, pedagogies, and theories to account for rhetoric's ongoing movement, performativity, and affectivity" (8). Instead of approaching a song, for example, as an isolated object—a thing to be interpreted—it is a process in motion. Sometimes this movement is deliberate (vs. tacit) in musicians' collaborative attempts to "make a song together," but, even then, the movement is beyond any one participant's control. In "Recovering Delivery for Digital Rhetoric, James Porter helpfully distinguishes between distribution and circulation, with distribution referring to an intentional attempt to distribute discourse and circulation referring to a discourse's potential to flow on its own accord across spaces and cultures (Gries 8). Our use of exchange includes both intentional attempts to collaborate, to distribute sonic elements among relatively small and large circles of artists and songwriters, and the more dynamic circulations where sonic elements are taken up, modified, and moved by both human and nonhuman participants.

As processes embedded in dominant cultural arrangements, exchange on some level is always shaped by power relationships among participants. The sonic rhetorical exchange practices we highlight and examine here intend to benefit all participants and support shared composition and performance goals, yet not all participants benefit in the same way. For example, Michelle's participants, after recording their soundscapes and capturing elements of the land, returned to the workshop space and listened to each other's recording in an informal exchange practice, where one participant's recording was offered in exchange for listening to all the recordings. Although each recording was played for the group, the give and take within processes of recognition and value were informed by technologies, bodies, and cultural ideologies. Some sounds, such as human footsteps through the hay on the barn floor, were easily identified (or located) and carried meaning-making potential, while other sounds, like the rustle of a human turning over in a hammock in the barn, contained less affective intensity and potential velocity. Later, Michelle collected those affectively and semiotically rich sounds and used audio software to remix (adjust the volume levels and sequence and layer the tracks) the tracks in order to communicate and archive elements of the land publicly. Although such exchange did constitute a collaborative process of soundscape creation, the give and take processes did not circulate equally or evenly among participants.

Songwriters exchange songs with other writers as a central practice in order to create or modify a song. In a common example, a "songwriters in the round" performance operates by trading off songs, where each performer plays and the others listen or even collaborate with an improvised backup instrument or a vocal harmony. Throughout the performance, the song's circulation is invoked as the song is modified or even transformed through spontaneous playing at that moment. Similarly, in a songwriters' group or informal "song circle," each writer offers a new song or a partial song, often with the explicit goal of receiving a response or notes on the writing. Not all participants benefit in the same way from such exchange, however. Some insist on their song being played in one exact way, while others embrace improvisational changes to their musical arrangement. Songwriters not experienced with collaborative playing or jam sessions can shut down completely if others accompany them. Some songwriters simply want experience playing in front of other musicians and resist getting any feedback. Song exchange practices are inherently complex and ephemeral, then, because the song at any moment exists within the exchanges possible in that location—those places where musicians come together to play music, record, and perform. These processes of exchange, however uneven, are still at the heart of sonic rhetorical circulation.

Our argument in this book is that sound is, first and foremost, grounded in cultural ways of listening, and sonic meanings emerge most clearly when we receive and resonate sounds with ourselves, amplify what we decide needs emphasis, often in collaboration with others, capture recorded sounds, and then exchange (e.g., share, reflect, remix) the sounds with participant others. We carry the resonant listening practices outlined above into chapters two, three, and four, with each chapter focusing on a different sonic location—from the routinized spaces of the classroom, to the regular rhythms of song circle, to the one-time Re/Call event. The chapters demonstrate in detail how all these sonic practices occur recursively and resonantly within particular classroom pedagogies, singer-songwriter events, and professional sound artworks.

2 Resonance and Culture in the Classroom

For many teachers, the sonic practices that locate us in classrooms and universities, that shape our sense of self, other, and belonging, often go unnoticed, even when the focus of the course is sonic rhetoric or sound studies. What also goes unnoticed, then, is how these practices are grounded in particular cultural ways of listening. This chapter details the resonant listening practices we taught and experienced in several of our courses, practices that exposed the unexamined cultural norms underlying the very sounds we detect and the ways we make sense of those sounds. The pedagogical and research methods we employed in our classrooms informed and were informed by the locative resonant listening practices we observed in our artist/musician collaborations—that is, the practices of receiving, amplifying, recording, and exchanging. We begin with classroom and university spaces because that is where our own work began but also because the sonic practices in those spaces are more routine for our readers. We hope to problematize and interrupt those routines before moving on to analyze the less familiar, more event-based sonic environments. Our resonant listening practices, which rely heavily on experimental sonic artworks and cultural soundscapes, move students beyond a static notion of sound as a singular, ear-based event and into more everyday listener-centric experiences. These experiences allow us to resonate more intimately with other active sound producers and amplifiers in the environment, such as rocks, water, air, bridges, buildings, mechanical engines, and non-human animals. Our practices also encourage students to shift from analyzing and composing with sounds as "objects" to analyzing and composing with sounds as part of a larger sociocultural soundscape.

While many of our students have a great deal of experience composing multimodal texts, they also remain relatively dissatisfied with their attempts to control and manipulate sound, whether it is an oral interview, soundtrack, or sound effect. Sounds do not stand still. Sound theorist and artist Salome Voeglin argues that sound renders objects dynamic in their ghostly and breeze-like surrounding of those objects (25). She further explains that sound, when "listened to *generatively*, does not describe a place or an object, nor is it a place or an object, it is neither adjective, nor noun. It is to *be in motion, to produce* . . . In this appreciation of verb-ness the listener confirms the reciprocity of his [sic] active engagement and the trembling life of the world can be heard" (30–31; our emphasis). Like Voeglin, we advocate a generative and resonant form of listening to sound as a verb. Similarly, Selfe emphasizes both the "movement" and "breath" of meaning in sound so as to "encourage teachers to develop an increasingly thoughtful understanding" of aurality as a historically and culturally powerful "semiotic resource" (618).

Before detailing our resonant-based approach and classroom experiences, we want to chart our own shifting understanding of sonic pedagogy. We begin with Kathleen Yancey, who includes audio in her widely quoted 2004 article, "Made Not Only in Words: Composition in a New Key":

> [W]riting IS "words on paper," composed on the page with a pen or pencil by students who write words on paper, yes but who also compose words and images and create audio files on Web logs (blogs), in word processors, with video editors and Web editors and in e-mail and on presentation software and in instant messaging and on listservs and on bulletin boards—and no doubt in whatever genre will emerge in the next ten minutes. (298).

Yancey characterizes writing as a series of dynamic practices using multiple semiotic modes, including sound, but since that time, the early lists of modes that included "audio files" and "sonic literacy" in compositional process have evolved into detailed articulations of multimodal composition processes. While Yancey legitimized our interest in sound as a key compositional mode in the classroom, *Writing New Media* (also published in 2004) offered us the best theoretical foundations for analyzing sound-based media, alongside pedagogical approaches for teaching and assessing them. We, along with other rhetoric and composition scholars, used these foundations and approaches, which were inspired by the New London Group, Günther Kress and Theo Van Leeuwen, alongside Cope and Kalantzis, to theorize sound as one among many legitimate modes of composition and meaning making. Their research also inspired us to critique the affordances and limitations of specific sound technologies. Such early work in sonic rhetoric included the 1999 special issue of *enculturation* on "Writing/Music/Culture" edited by Rickert; the 2006 print and online companion issues on "Sound In/As Composition" in *Computers and Composition* edited by Cheryl Ball and Hawk; the 2011 special issue of *Currents in Electronic Literacy* on sound edited by Diane Davis; and the 2013 online journal *Harlot of the Arts* special issue of creative works on Sonic Rhetorics, edited by Stone and Ceraso. These groundbreaking publications included and also prompted more analytic and creative works in sonic rhetorics (for example, see Jennifer Bowie, Ceraso 2014, Elbow, Halbritter 2013, Jeff Rice, Rickert, Jentery Sayers 2012, Selfe 2009, and Stone 20).

Also in 2004, the first dissertation in the field of computers and composition on sound was written by Halbritter, which then laid the groundwork for a critical mass of dissertations on sonic rhetoric like those completed by Kati Fargo Ahern (2012), Ceraso (2013), Sayers (2011), Kyle Stedman (2012), Stone (2015), and Crystal VanKooten (2014), among others. Along with these scholars, we moved away from researching and teaching sound as supplementary to print and visual texts (e.g., read-aloud revision strategies, think-aloud protocols, soundtracks, scripted voice-overs, etc.). For example, in 2006 we defined sonic literacy as a multisensory practice—as "the ability to identify, define, situate, construct, manipulate, and communicate our personal and cultural soundscapes"—and as "a critical process of listening to and creating embodied knowledge" (Comstock & Hocks, "Voice in the Cultural Soundscape"). By this point, we had already discovered composing practices used by sound artists Pierre Schaeffer, Chion, and Susan Philipsz that served to liberate voice-overs, soundscapes (ambient and environmental sounds), and soundtracks from the work of non-diegetic narration and exposition. As Philipsz makes explicit even in her earliest installations; all sounds, including voices, are material multi-tonal dynamic objects in their own right, worthy and capable of close observation and modest manipulation (Pollock). Once we began understanding sounds as material objects or, in the words of Steven Goodman, as "vibrational surfaces, or oscillators" (7), which carry both semiotic and non-semiotic messages about experience and the environment, we moved from characterizing sonic rhetoric as a mode of discursive persuasion to defining it more broadly as an embodied and dynamic engagement with sound.

Often human engagement with sound is tacit. It plays an autonomic, yet powerful, role in creating and locating us within what R. Murray Schafer calls "the surround," a term he coined in his 1977 foundational work *The Soundscape*. Much of his work then and now has been dedicated to cultivating more conscious awareness of "the surround," though in Schafer's case it is a white settler colonial soundscape protected from "noise" of urbanization and Indigenous forms of music and chant. You'll see in our assignment sequence that Schafer's terms—"the surround" and "the soundscape"—still do a great deal of conceptual, cultural, and ethical work, though we have shifted the emphasis from colonial notions of the pristine unsullied soundscape to soundscaping—a highly rhetorical and cultural locative sonic practice instead of an actual place or thing to be discovered. Rickert's work has helped the concept of sonic rhetoric move forward in this regard. Like Schafer, he argues that the ambient noises of our daily experience often go unnoticed, but he stresses that these ambient noises are part of dynamic rhetorical situations, situations that continually construct our very sense of self and the world (20). Thus, our pedagogy begins with practices in receiving and amplifying ambient noises in order to identify the ways they shape our sense of self, other, and the environment. These resonant listening practices can be difficult because they require a particular vulnerability and sensibility to the sonic environment, as well as a great deal of dedicated and sustained attention to layers of sound, to how those sounds resonate within certain parts of our bodies, and to the very ways we choose to pay attention and filter information.

Marilyn Cooper's work has also expanded our definition of sonic rhetoric from simply a persuasive element or affective layer within a multimedia project toward a vibrational interaction with a complex environmental system. Her model of rhetorical agency is particularly helpful to our resonant listening practices as she describes individuals as agents *only* within complex systems:

> Complex systems (an organism, a matter of concern) are self-organizing: order (and change) results from an ongoing process in which a multitude of agents interact frequently and in which the results of interactions feedback into the process. Emergent properties (such as agency) are not epiphenomena, nor "possessions" in any sense, but function as part of the systems in which they originate. And causation in complex systems is nonlinear: change arises not as the effect of a discrete cause, but from the dance of perturbation and response as agents interact. (Cooper 421)

This "dance of perturbation and response," we suggest, helps to characterize an individual's voice, for example, as merely one resonant dynamic among many within a cultural soundscape. Even the human voice itself can vibrate at multiple tones (overtones) within the same frequency, thus offering multiple potential points of resonance. We, therefore, understand the artist-composer as a vibrating subject and object—a subject-object that is imbued neurobiologically with the ability to resonate and empathize with multiple sounds, tones, and frequencies within an environment, and who is also enculturated in certain ways of listening depending upon positionality and training. Cooper's notion of positionality—as dynamic subjects moving along trajectories within the surround and the environment—allows for the complex forms of intersubjectivity and agency a sonic rhetoric requires. Distinct subjects and objects do not make sense in relation to sound, which is inherently vibrational with multiple waveforms constantly arising and decaying. Instead, following Cooper, sonic rhetorical positionality becomes a trajectory or path.

During the sonic composing process, the technologies themselves and their visual representations within the digital interface also participate in the "dance of perturbation and response" and play a role in moving us along particular cultural, historical, and political trajectories. Every sonic representation is constrained by its interface and primary interpretative method. A sonogram or an MRI, to name a high-stakes medical example, is structured by the questions and methods of medical diagnosis, and interpretations are often based on assumptions about culturally normative human bodies. In contrast, the waveforms and multi-track mixers used for a professional Digital Audio Workstation (DAW) and for personal audio composing software allow composers to see and shape the constituent features of a sound, such as pitch, volume, and decay. Furthermore, multi-track mixers allow composers to divide sounds into individual tracks and see and move them in relation to each other. Thus, digital technologies—from microphones to mixers to file compressors and speakers—all expand our sense of rhetorical sonic agency.

We advocate teaching sonic rhetoric as a dynamic interactive practice, which involves listening carefully to the ways technology and the environment shape what we hear, amplify, record, and exchange. When students adopt this resonant listening approach, they move away from composing practices aimed at producing stable, trained, third-person voice-over tracks (tracks popularized by mainstream radio stations, political advertisements, and nature documentaries) and toward composing practices that integrate the human voice with other active sound producers and amplifiers in the environment. The processes and pedagogies described here demonstrate to students that the resonant listening practices of receiving, amplifying, recording, and exchanging are experienced as:

1. completely embodied, multisensory, and vibrational;
2. applicable within any sonic environment—urban, industrial, rural, or virtual;
3. mediated by cultures, spaces, and technologies; and
4. enhanced by reduced listening, by tuning into particular sounds and their dynamic physical qualities.

Such an approach requires students to analyze the intricate nature of sound objects, along with the dynamics of how we experience them, before they begin to compose their own sonic texts and multimodal projects.

Resonance as a Material, Cultural, and Rhetorical Process

Physically, resonance refers to the impact of one vibration on another and the ringing of multiple sympathetic frequencies or tones in (usually) harmonious ways. Metaphorically, we often use the term for a wide range of meaning that indicates our harmony and connection with a text, a place, an idea, or an object. In our sonic pedagogy, however, resonance is an umbrella term for the intimacy, presence, and movement (the "verb-ness") created by a sound's qualities, like tonality, amplitude, or cadence. An everyday literal example of resonance is hearing a paper bag rustling in the next room. We hear (or do not hear) the rustling bag because our inner ear vibrates (or does not vibrate) at the same frequency. That resonance with the paper bag evokes immediate experiences and sensations, as well as memories and meanings. Most sounds in this world, of course, are actually outside of our physical range of hearing and/or our individual ability to resonate. However, even acknowledging our limited range of resonance gives agency and power to the sounds outside our conscious awareness. For example, the National Park Service (NPS) explains the conservationist work of acoustic technicians conducting field assessments and adjustments with all elements of the "acoustical environment" regardless of human accessibility:

> The sound vibrations made by our imaginary falling tree are a part of the acoustic environment regardless of whether a human is there to perceive them. Bat echolocation calls, while outside of the realm of the human soundscape, are also part of the acoustical environment . . . Because the acoustic environment is made up of many sounds, the way we experience the acoustic environment depends on interactions between the frequencies and amplitudes of all the sounds. Sound levels are often reduced or adjusted ("A-weighted") to match the hearing abilities of a human or given animal.

Because healthy ecosystems depend upon acknowledging interactions among *all* sounds within an acoustic environment, our sonic rhetorical approach takes into account how a listener's auditory system, as well as the shape of a particular space, shape their ability to vibrate at particular frequencies over others. Further, this approach asks whether the listener can generate empathetic tones with others within a larger sonic field. We hedge with "can" here because we emphasize the dynamics of empathy and affect while being acutely aware of how our experience of sound is dependent upon biology, environment, evolutionary sensory processing, and culture. Even our complex biological sound processing systems are still largely theoretical descriptions of signal and noise that link physiological mechanics in the ear with the brain's mysterious processes. Scientists and medical doctors still readily admit they know little about how an individual's auditory system separates "signal" from "noise" (McCarthy). Further, physiological absolutes and sensory metaphors rarely account for the lived experiences of an individual or for differently abled bodies within and across different cultures.

We want to emphasize that resonance is not limited to hearing-centric and ableist definitions of listening. Sound studies scholar Jonathan Sterne writes in *The Sound Studies Reader* that deafness has always been closely involved in research in sound technologies: "[T]he deaf and hard of hearing are everywhere in sound history, both as objects and subjects . . . [I]t is not just a question of the limits of hearing for those who are deaf or hard of hearing, but rather that sonic thought needs to move toward the edges of sound itself—and beyond—to consider sound as a particular moment in a broader ontological and political field of vibration" (20). People who are hearing impaired listen in a multitude of ways, including but not limited to their perceptions of vibrations in other parts of their bodies. As Ceraso argues, "[I]dentifying the ear as the body part that enables listening does not capture all that is involved in experiencing a sonic event. Listening is a multisensory act" (102). Based on her interview with deaf percussionist Dame Evelyn Glennie, Ceraso introduces the concept of "multimodal listening" in order to expand how we think about and practice listening as a situated, full-bodied act" (103). Scholars working at the intersections of digital writing and disability studies (for example, in the special issue of the journal *Kairos* on "Multimodality in Motion") emphasize the lacunae in our discussions about multimodal composing and the effects of marginalizing different experiences of hearing and listening. For someone who is deaf or hard of hearing, resonance and vibrations are experienced differently, for example by body language, sensing vibrations in the body, and lip reading. Therefore, listening must be understood as more than just an ear-based "fact"; instead, it is an act of engaging the world.

Resonance also depends upon cultural norms, disciplinary training, and sonic traditions. As Robinson argues, "[L]istening is guided by positionality as an intersection of perceptual habit, ability, and bias" (37). Settler positionality, in particular, guides forms of listening that privilege conceptual information at the expense of a sound's "feel, timbre, touch, and texture" (Robinson 38). In terms of disciplinary training, musicians are trained in an aesthetic form of listening, and they learn to resonate with certain sounds over others. Often practice includes imitating or singing and playing these elements until they "sound" right. That "sounding right" is the moment where the trained musician or experienced listener resonates with his or her own experience of a specific musical tradition. This cultural tradition is broader than just taste, genre, style, though those elements certainly come into play. One student in Mary's class, for example, wrinkled his nose every time an acoustic guitar played because he heard it all as "country" music. He preferred bass-driven improvisational jazz. Of course, often the most innovative and interesting music breaks cultural rules and blends styles. Whatever the tradition, that historical moment's dominant cultural arrangements and artistic styles will play a large part in how one hears and resonates with music. Resonance with music (something with which most people can identify easily) is culturally based and fully embedded in a socio-historical context, which is why our pedagogy is carefully sequenced to move students from experiencing sounds as extractable objects to understanding listening itself as a cultural, rhetorical, and material process.

We also teach resonant listening practices because certain forms of resonance, if cultivated, may offer writers stronger positions that are more responsive to current changes, for example, in our ecosystems, daily individual rhythms, and political and social landscapes. Resonance also prompts us to explore the relationship between sound and written language. For example, we ask ourselves and our students what happens when we read our writing out loud, give breath and vibration to our words, and project them. Is writing really, as Voeglin suggests, a "quasi process of speaking" (48)? Is intoning the sounds of words a way to improve writing, as Elbow suggests in *Vernacular Eloquence*? How is writing more than, less than, or different from speaking? Does focusing on sound modalities and tuning into the multidimensional processes of speaking and listening automatically make us better communicators and writers? In other words, what happens when we write *for* a resonant voice? And how might this activity make us more responsive to our immediate environments, including our own bodies? These and other questions led us to an extensive and flexible semester-long assignment sequence built on the resonant listening practices we developed in concert with artists and musicians–receiving, amplifying, recording, and exchanging. The sequence should be helpful for instructors who are both experienced and new to multimodal composition or sonic rhetoric.

For more than a decade, we have cultivated resonance in our classrooms by moving students through a sequence of audio assignments that rely on phenomenological and cultural forms of listening and composing. Each element in the sequence emphasizes the deliberate aesthetic and ethical practices and gestures students can make as they learn to listen and compose. We first ask students to engage in daily phenomenological reduced listening and writing practices using sound art and music as their objects, which helps them become more receptive to subtle and diverse sonic influences and fosters an ability to amplify—both internally and externally—marginalized and ambient sounds and voices. These reduced listening assignments are followed by a sonic memoir assignment that asks them to record and remix sound clips to frame a listening self and to reflect on how our very sense of self is informed by sound. Students then end the courses by researching and documenting their listening/resonating selves within larger sociocultural soundscapes. The sound documentaries become part of an exchange process within the course, as students use them to sonically map a larger region. To summarize this sequence in phenomenological terms, students move from resonating with sounds as objects to resonating with themselves as listening subjects to resonating with a larger cultural sonic field, where the sound object and vibrating subject become more diffuse and interdependent. While the sequence does provide a scaffold for training the receptive body first before analyzing and documenting the larger sonic field, it would be a mistake to ascribe a strict concrete to abstract scale to the assignments. Instead, students notice what happens to meaning-making processes as they shift their attention from a sound object to sonic interactions within a field.

While following the overall sequence, which we developed in parallel at two universities over several years, each of us modified assignments to meet course and university learning outcomes. We include a number of assignment options for each part in the sequence, along with emphasized resonant listening practices, in the following two sections:

- Resonating with Sound Objects, Reduced Listening, and Framing a Listening Self—Sound Art, Music, and Sonic Memoir Assignments—Receiving, Amplifying, and Recording Practices
- Documenting Cultural Soundscapes—Audio-only Documentary and Multimodal Assignments—Exchanging Practices

Although the sonic practices of receiving and amplifying, for example, are emphasized in the opening assignment, they recur throughout all the assignments, along with a focus on the cultural norms and positions that guide such practices. Students studying and composing with sound need to move their attention from sound objects to the very ways they perceive and make sense of sounds in order to transform dominant and harmful ways of listening to themselves and others within classroom, community, and environmental spaces.

Resonating with Sound Objects, Reducing Listening, and Framing a Listening Self

Originally adopted by Schaeffer as a way to access the essence of sound through recordings (what he called "acousmatic reductions" because of their distance from the source), "reduced listening" was a technique that separated the listener from the sound's notation, means of production, and the listener's own mind state (Demers 26). Of course, this is an idealized process, but it can interrupt what Schaeffer identifies as our four common modes of listening: "ecouter," which focuses on the qualities of the sound that we find appealing; "ouir," which marks the physiological process of listening; "entendre," which highlights particular aspects of the sound that we seek to understand; and "comprendre," which refers to the ways a listener engages with the sound via its external references or its "context of associations" (Demers 29). Reduced listening seeks to "bracket" or pause these common forms of listening through the removal of visual cues and sonic sources, so the listener gains insight into their own perceptions, develops a receptivity to a wider range of sounds and sonic environments, and begins to hear in a new way.

Instead of bracketing associations, however, Schaeffer's student Chion advocates attending to those associations in order to see how our perceptions and interpretations are constructed and are inextricable from our bodies and environment. Chion divides listening into three (vs. Schaeffer's four) modes: causal, semantic, and reduced. In causal listening, the most common form, one listens for information about the source (sound is supplementary), while semantic listening interprets for code or meaning. "Reduced listening," as defined by Chion, is the conscious bracketing of the first two modes in order to make the sound itself—its physical characteristics, such as its point of onset, pitch, frequency, volume, timbre, duration, and collapse—the object of attention. He acknowledges that the difficulty (if not the impossibility) of this practice is, in fact, the point. When we engage in reduced listening, we grasp for source and meaning and often find our language for describing the material effects of sounds impoverished (50). Chion's model is extremely helpful when teaching students how to explicate sounds that are emotionally charged with an immediate reference or connection to a narrative memory or sounds that carry no direct reference or any explainable meaning to the listener. Such listening, which invites students to resonate with (vs. interpret) emotionally charged as well as neutral sound objects, takes practice, and as Chion notes, we defend against it through either measuring the sound in some way (with a stopwatch, for example) or by claiming subjective relativism (e.g., "everyone hears something different"). Chion argues that there is, in fact, a kind of "intersubjective objectivity" that situates our own listening practice alongside those of other listeners and prevents any pure individual phenomenon (50). To put this another way, because listening relies on perception and individual perception is based on learned "shared perception," all listening, according to Chion, is intersubjective (50). For example, the sound of a dog barking arises in an individual's mind because she has been socially primed and conditioned to isolate and perceive that sound against an ambient field of other potential sounds.

Why is reduced listening beneficial to students, particularly if they do not plan to become sound artists or engineers? Is it enough to say that reduced listening convinces them of the power (and affordances) of sound itself? Chion argues, "The emotional, physical, and aesthetic value of a sound is linked not only to the causal explanation we attribute to it but also to its own qualities of timbre and texture, to its own personal vibration" (51). These vibrations, with their kinesthetic as well as sonic qualities, are not necessarily "more true" or "more real" than images, but they do have rhetorical power. Even oscillating between reductive listening and causal listening can bring more awareness to the qualities of the sound and its effects, especially its ability to (dis)locate us emotionally and socially. Furthermore, it can bring attention to the limits of conscious attention. For many of us, our capacity to filter and select sounds is far more limited than our ability to sort and prioritize images. It is only upon reflection that we can notice the ways particular sounds emerge as salient from what Rickert calls the "ambient dimensions" ("that which escapes conscious notice") of any rhetorical situation (10).

Schaeffer and Chion's own recordings, which amplify the "noises" of industrialization and commercialization, offer in-class opportunities for collective forms of reduced listening. As we listen together to Schaeffer's "Etude aux Chemins de Fer" and Chion's "La Ronde" and note each flux in sound and overall texture, we develop a practice that makes us more receptive to and aware of (with awareness as one form of amplification) human and nonhuman elements vibrating interdependently in a particular place and time. Although it would be problematic if we mistake these reduced listening practices as inherently more natural or essential, Chion's techniques have helped students cultivate the ability to receive and amplify a diverse set of ordinary ambient sound objects in their environments.

Assignment: Listening to and Resonating with Sound Objects

Ask students to listen to Schaeffer's "Etude aux Chemins de Fer"—his first *musique concrete* recording—and Chion's "La Ronde." Prompt them to first listen causally, then semantically, then ask them to bracket both, so they hear the qualities of the sound itself. Ideally, you want students to use their own words to describe the physical qualities, but if they are having trouble, prompt them with general terms like timbre, rhythm, frequency, and duration. Ask them to listen as a class, through a group meditative listening, and then listen individually through headphones, which provides more sensory deprivation and allows them to focus but also monitor where their attention goes. Next, ask them to perform reduced listening on a recorded sound object outside of class. Finally, ask them to record a "found sound" object for the class. The class listens causally, semantically, then brackets those two and practices a reduced form of listening. Through journal writing, the class can then reflect on their listening experiences and their difficulty or ease in finding language to describe them. If students are having trouble articulating their experience, we use the following prompts:

1. Describe the birth, growth, peak, decline, and end of the sound object, as well as its repetition or rhythm. Try to bracket your impulse to identify the sources and meanings of the sound, so you can understand its physical qualities. Since completely bracketing our meaning-making impulses is difficult, if not impossible, observe how those processes guide your perceptions and experiences of sound.

2. Next, lighten your focus on the sound object and notice how you are reacting to and making sense of the objects in parts of your body and mind. Sound artist Pauline Oliveros asks us to notice where we feel the sound in our bodies and what our physical sensations are when we hear the sound. Finally, note the "feeling quality" (e.g., pleasant, unpleasant, pleasurable, painful, joyful, sorrowful, etc.) of your awareness of each sound.

Assignment: Training in Analyzing Music with Reduced Listening Techniques

While using sound art to train our students in reduced listening, we also listened to musical recordings (with no video) in order to cultivate a sense of resonance in the classroom. These reduced listening practices, when combined with focused attention on experiences and affective responses to sound, create opportunities for students to respond immediately to music and to focus more on recreating their immediate experiences of a sound before moving into an analysis of those sounds. To formalize this analytic process for students, a sequence of reduced listening practices focuses on recorded songs to encourage experiential listening and analytic responses rather than simple taste. Ask students to provide a link (*without video*) or a sound file for the specific song and then answer structured questions in this order:

1. Do you like this song? Why or why not? Listen to it again, paying careful attention, and make some notes in sequence to try and recreate your actual listening experience, using what we have discussed as a phenomenological method of reduced listening.
2. When did you resonate with or feel connections to this song? When did you feel dissonant disconnections? When did you simply stop listening and tune it out? Try to pinpoint the moments in the song where these experiences happen.
3. How would you describe your intellectual reaction (lyrics, meanings, memories)? What does it remind you of? Write down words that you associate with it.
4. Did hearing other people's songs in class change your approach to your own song? Why or why not?

Assignment: Framing a Listening/Resonating Self through a Sonic Memoir Playlist

After experimenting with reduced listening and reflecting on the ways we receive, amplify, and make sense of sound and music, especially in relation to timbre, frequency, rhythm, and duration; we move students toward creating a stronger sense of the listening self and body through a sonic memoir playlist assignment. Students are expected to use sounds metaphorically and thematically in order to tell their classmates something about themselves. We borrowed Sayers' "audiography" assignment, with some minor modifications, to explore (without relying on voice-over) how our sense of self, including our experiences of belonging and alienation, are informed by sound. Students record and remix sounds into a playlist that leaves listeners with a strong impression of who they are at this juncture in their lives. For this assignment students begin engaging with recording practices—field recording, in particular—and, therefore, need some kind of microphone (built into a phone or gaming headset or via a USB external microphone) to record ambient sounds and noise. We refer them to Halbritter's chapter on microphones, which provides detailed, useful information on how to make these choices (117–61). The point, though, is not to find or purchase an expensive microphone that blocks out "noise." Instead, it is to experiment with and understand how sounds are co-created with technologies, including grainy smartphone microphones and homemade contact mics. Distortion and noise become part of the material and, ultimately, part of the composition. We have found Steven Hammer's essay "Writing Dirt, Teaching Noise" helpful in orienting students toward a deeper understanding of noise as an essential and strategic part of sound composition. In fact, one of Michelle's students, who worked at the airport, built his sequence on the shades of white noise—some more intense than others—that he experienced through "protective" headphones on his job site. Like Schaeffer and Chion, he created a rhythm out of the muffled drone of low frequency engines to convey his body's movement among them. Also, using Hammer's definition of noise (i.e., "Noise makes us work harder"), he conveyed the additional labor it takes to both process and ignore the sounds of his workplace.

One of the difficult objectives of this assignment is to communicate an overall theme or metaphor to the audience, which puts considerable semiotic weight on not just the selection and recording of sounds, but also on remixing and sequencing them. Students work directly with pacing, rhythm, vibration, and resonance, but this time without their usual reliance on speech or writing. Several of our students framed their memoirs within seasons or seasonal changes—an apt choice for the medium, given that our experiences of climate (and the social events associated with changes in climate) are strongly informed by sound. A snowstorm, for example, provided a context for contrasting the morning sounds muffled and slowed by the recent snowfall and the crisper, busier sounds of the afternoon thaw. Another student cataloged and remixed the sounds of her annual Halloween ritual. The memoirs ultimately do more than simply profile individuals; they tell us how students listen to their soundscapes and how those soundscapes (dis) locate them in particular bodies and identities. Sayers himself later notes in his "Audio Culture Series" the difficulty of this assignment because it is not rooted in any particular genre other than the general playlist genre; thus, students must rely on the highly technical directions for audio remixing more than any model texts or samples. However, sound art, once again, provides a good resource for the sonic memoir. For example, students listened to modern versions of *musique concrete* for forms of memoir composed through a carefully remixed sequence of sounds. Again, Chion's "La Ronde," to name one example, represents his nostalgic experience of the sounds of a seasonal urban circus, and the more contemporary "Tongues of Fire" by Trevor Wishart shows students how distortion and noise can powerfully communicate emotional psychological experience. Even with the art samples, students do find the assignment challenging. It is something they have never done before, and they often resort to the genres they do know (e.g., the mixed tape or a radio play). Still, through the experience of recording and remixing with a multi-track interface, they gain a deeper understanding of the affordances and limits of working with sound.

Documenting Cultural Soundscapes

After exploring the contours of the hearing self-body, students shift to analyzing and documenting their cultural soundscapes, beginning with their immediate campus contexts. Arguing that "the general acoustic environment of society can be read as an indicator of social conditions which produce it," Schafer offered language for this important rhetorical and epistemological move from subjective to cultural analysis (7). His key terms for marking various soundscapes are part of a larger sonography, which includes the "*keynote* sound" (a background or fundamental tone against which other sounds are perceived, such as the sea), the *signal* sound (the foregrounded sound to which the attention is particularly directed, such as a boat whistle), and the *soundmark* (a community sound that is unique or possesses qualities that make it noticed by people in that community, such as church bells) (9–10). In first analyzing their campus sounds more closely, students looked at their physical characteristics using technical terms, such as the *envelope* (the attack, the body, the transient changes, and decay); the *psychoacoustics* (or the way the sounds are perceived); the sounds' emotional or affective qualities; and the sounds' role in locating individuals within social and cultural contexts. Employing Schafer's heuristics, students then assessed how institutional and cultural forces determine what qualifies as a viable sound, voice, or listening self/body-mind.

Since we first began assigning soundscapes in our composition courses in the early 2000s, rhetoric and composition scholars have developed increasingly sophisticated approaches to soundscape analysis and design. Ahern, for example, offers a new multidimensional heuristic that helps students analyze and design soundscapes for classroom and professional audiences. Her four dimensions are 1) the source of individual sounds; 2) "[t]he temporal unfolding of sounds and sonic events among, against, and with each other (for example, the unfolding of a rain storm);" 3) "[t]he multiple layers of simultaneous sound;" and 4) "[t]he location of sound sources in space, the space of the soundscape, or the spatialization of sound (forming relationships such as perspective and movement, as well as location in the space)." Ahern's heuristic, which is highly recursive and complex, puts students in the mindset of a sound designer and as active participants in the construction of soundscapes inside and outside the classroom. They grow to see that acoustic decisions were made—whether consciously or not—and they begin to understand the impact of these decisions on human and nonhuman bodies. For rhetoric scholars and instructors, Ahern helpfully connects each dimension to a rhetorical concept, creating what she calls a productive "entwinement" between each disciplinary concept: 1) sound sources relate to *doxa*; 2) temporal unfolding to *dispositio*; 3) simultaneous layering to *copia*; and 4) spatialization with *saphes*.

As Ahern demonstrates, multiple translations and interpretations of each Greek term have been provided by scholars in the field, which are useful for students to explore as they produce and design their spaces. *Doxa*, for example, can simply mean common belief or it can also mean that "intersubjective objectivity" Chion refers to in his discussion of reduced listening. Ahern is most interested in the term's usefulness for sound design and production. Thus, *Doxa*, depending on how one defines and utilizes it, can be a productive place to start or can stymie creativity. *Dispositio*, often understood as arrangement, can connote sequencing, ordering, pacing, and/or tempo; *copia* can suggest multiplicity (stating the same message in different ways), amplification, and layering; and *saphes* can mean clarity, conciseness, precision, and/or locatedness. Such a rich interdisciplinary heuristic provides instructors in both rhetoric and acoustics with additional analytical and design tools that help students better understand the persuasive power of sound within bodies and environments.

Assignment: Campus Soundscape Analysis

This first assignment uses the immediate campus soundscape to invite students to analyze the ways certain sounds regulate movements, gestures, and identities within particular spaces. Limiting the exercises to two class periods serves to isolate the listening and analyzing practices from other everyday interruptions and distractions, and it provides collective (group) support for the difficult attention work. Ask students to create groups and record what they consider the keynote, signal, and soundmark sounds on campus. When they return, ask them to write responses to the following reflective prompts:

1. Specific moment questions: Select a particular moment you experienced in relation to the campus soundscape. What did you have to do in order to listen to what was happening? Did you experience a sense of resonance with the sounds? How did you know you were resonating? If you did not experience resonance, what did you experience or when did you disconnect with the sounds?
2. Cultural and political questions: What does the campus soundscape tell us about our culture(s)? What is the sonic history of the cam-

pus? What voices/sounds are privileged and what voices are silenced? How do the sounds in your chosen soundscape inform participants' relationship to time? To their environment? What kinds of sound walls are constructed to block out unwanted sounds? What kinds of relationships do the sounds create?

After reflecting on their experiences listening and recording, students will then play their recordings for the entire class. During and after this exchange, they can work together to compose (or at least outline) an overall campus soundscape that will communicate to both new and returning students that they are on a particular campus. Through this exchange and composing practice, certain sounds are picked up and assembled over others, and students should talk about why this is the case. Sometimes it has to do with fidelity and/or the quality of the sound; more often, it relates to the metaphorical qualities of the sound—for example, heavy doors opening can suggest social, cultural, or institutional doors opening (or closing, for that matter). The sound of a wheelchair entering the building can also spark cultural notions of access or lack of access. The dynamics of "noise," "sounds," and "silences" also become cogent cultural markers for students in this process. Many students, for example, create a sequence that moves the listener from the so-called noise of car commuting, including the mixed layered frequencies of car engines and playlists, to the relative silence of a classroom building. That shift is what delineates campus for them. To problematize this cultural assumption a bit, we ask them to record a place on campus that is more polyrhythmic and multi-tonal—perhaps a library, a gym, or a gallery. How do these spaces inform the commuter campus or commuter student identities that our students are so quick to recognize or construct? After composing and exchanging the campus soundscape analyses in class, students move to recording and analyzing soundscapes from their everyday lives.

Assignment: Environmental Soundscape Analysis

This second assignment asks students to expand their campus analyses to another soundscape in their everyday lives. Ask students to record an outdoor soundscape they regularly encounter and reflect on these questions:

1. What kinds of environmental sounds did you respond to within this soundscape? Was it an urban soundscape, a natural soundscape, or a combination? Listen again to your recording and describe in writing your listening experience step by step.

2. Do you like this soundscape? Why or why not? When did you resonate and feel connections to this soundscape? When did you feel disconnections? When did you stop listening or tune it out?
3. How do you feel when you hear this soundscape? How would you describe your emotional reactions (annoyed, happy, peaceful, connected or disconnected, etc.)? Your physical reactions (heart speeds up, tenser, relaxed, peaceful, grimacing, hands over ears, moving slower, moving faster, etc.)?
4. What do you recognize in this soundscape? What does it remind you of? Write down words that you associate with it and experiences or stories that it reminds you of.

Students tended to choose familiar spaces—from churches to parking lots to favorite tattoo parlors—that they could then re-experience using the phenomenological listening techniques that the prompting questions require.

The above short recording and analytical exercises then culminate in a final major assignment that uses a recorded soundtrack, podcast, or website to move students from analyzing to documenting a particular soundscape or sonic event. Equipped with both a subjective and cultural sense of resonance, students could adopt a narrative, argumentative, expository, and/or experimental structure to compose (bring various elements together) and exchange their chosen soundscapes with larger audiences.

Since first assigning soundscapes, we have watched a particular subgenre grow in popularity among artists and students—the soundwalk or audio tour. This subgenre can lend itself to a narrative structure (a journey through a forest or neighborhood) but not necessarily a linear one. Because it includes the "walker" or "experiencer" of the soundscape as part of the composition, it focuses on the human-environment interactions and interdependencies and makes the cultural assumptions and values informing these interactions more explicit. One of the more famous soundwalks is Hildegard Westerkamp's "Kits Beach Soundwalk," which offers Westerkamp's stream-of-consciousness reflections on the sounds in her immediate natural surroundings. She makes sonic space for the impacts these sounds have on her values, dreams, and culture and vice versa. The human voice itself and her footsteps through the beach become an integral part of the soundscape and emphasize the growing human influences on the planet in general. After students listen to and analyze Westerkamp, we ask them to listen to another famous soundwalk, Janet Cardiff's "Wanås Walk." Cardiff, like Westerkamp, includes the human voice to narrate dreams, tell stories, and reflect on the sounds and elements in the landscape. However, one of the human voices addresses listeners directly *in situ*, telling them where to walk and pointing out landmarks along the way. Though the voice isn't completely trustworthy (it gives way to other human voices shouting in the distance, singing, sharing dreams, etc.), it strongly mediates the listener/walker's experience of the surrounding woods. Many students end up combining the two approaches by creating soundwalks that can be accessed *in situ* on a smartphone (in a neighborhood or on a mountain hike) or from a desktop computer inside a dorm room. They use the human voice combined with other sound elements to create a sense of place (what Ahern also calls *saphes*) for the listener, as well as a sense of their individual interactions with the environment and technology.

One of Michelle's students, for example, recorded his regular two-mile walk to the university through the Capitol Hill neighborhood in Denver. Because it was a cityscape with plenty of human interaction, he recorded impromptu conversations with passersby about landmarks along the way. The rest of the soundwalk includes cultural (and not necessarily local) soundmarks that locate us on a children's school playground via the traditional sounds of children laughing and playing, a retail street with the traditional sounds of store music and "may I help you" questions from the workers, and a series of questions from people wanting change. Without the conversations or without experiencing the soundwalk *in situ*, listeners could have imagined themselves in a number of US towns or cities or simply taken the student's recordings as representative of his everyday sonic experience. At one level, that is all the assignment can do—that is, offer a brief snippet of a student's everyday sonic experience as mediated by a particular microphone on a particular day in a particular place. To do even that, however, students needed to rely on the textual discourse of the project title and their own names and/or the use of their own voices to locate us in the experience. Like the sonic memoir assignment, the soundscape assignment asked students to grapple with the sonic construction of a listener and experiencer, if not a self.

Assignment: Audio-Only Documentary

As we imply in the introduction to this chapter, voice-overs in our classrooms became objects of inquiry more than tools for sonic rhetoric. Students explored how voice-overs have historically made documentary soundscapes more or less "real" or how they have (dis)located us within particular environments. Before creating their own audio documentaries, students read Jonathan Kahana's *Intelligence Work* and modified his taxonomy in order to categorize and analyze several generic forms, including the professional examples of forms/styles listed below. They then noted how the different relationships between the subject and the narrator were mediated through the recorded speaking and/or singing voice. Voice-over quickly became a metaphor for the subject-object relationship, and not just a documentary technique or exercise in authorial voice. Here are key types of voice-overs and examples of each:

1. Voice of the State: Steiner and Van Dyke's "The City" from 1939
2. The Deictic Voice-Over: Dale Minor's "The Second Battle of Selma" from 1965
3. The Narrative Voice-Over: Maya Gurantz's "Christmas in 3-D"
4. 4. The Multiple Narrator Voice-Over: Dmae Roberts' "Mei Mei, A Daughter's Song"
5. 5. The Absent Voice-Over: Alan Hall's "William Thompson IV's War"
6. The Voice-within Soundscape: Philipsz' "Lowlands" (47 Film).

Blending the deictic and narrative categories, two of Michelle's students, Sonja Engelstad, and Kelsie McWilliams, produced a soundscape documentary on the campus Writing Center, where they worked as consultants at the time. Their goal was to record and narrate the keynote and soundmark sounds that locate us in the Writing Center. Even though their strong narrative voice-over nearly drowned out the deictic sonic material (the writing consultant's voice), the two tracks remained in tension and provided a "third," perhaps unintended, meaning. If we take the writing center consulting session as the primary sound object in this documentary, the voice-over attempts to categorize the object within a taxonomy of listening practices (i.e., polite, introductory listening followed by dialectical listening), but the attempt was not entirely successful. The session recording suggested the consultant (and one of the authors) was listening for evidence of learning, while the voice-over asserted a much more collective knowledge-making experience. Perhaps the voice-over strove to frame the session as an idealized dialogue, while the recording captured a messier, fairly directive, yet instructive, consultant-driven experience. "Capture" is a misnomer, of course, as the composers carefully selected and edited the session recording. The tension between the session dialogue and the voice-over beautifully expressed the mixed agenda at work in both the writing center session and in the documentary itself.

As teachers, we almost always first ask: How does one assess such a piece? Much of the assessment of the audio documentary occurred during the classroom exchange of the sound files. Students, many of whom already held strong personal versions of the Writing Center soundscape, suggested that Sonja and Kelsie revise to explore the tension between the two agendas and more fully attune their voice-over to the ambiguities of tone, rhythm, and emotional cues within the recorded session. The class also suggested the voice-over might reflect on its own role not only in the documentary but also in the Writing Center. For example, they could focus on questions about access and equity: What kinds of voices, tones, and rhythms regularly structure or dominate the soundscape of the Writing Center? Similar to the exchange practices in the next two chapters, these classroom exchanges contributed to a dynamic virtual sonic archive of the campus.

Assignment: Multimodal Blog Project

Another final assignment gives students a choice for a final project where they can further develop their soundscape analysis of a place and event and expand their research to define a current exigency related to sound-based research, debates, or language practices. For this assignment, students produce a blog with media—in this case, short original audio-visual clips or audio collage/remixes of found audio that illustrate the exigency of their topic. Students create short audio/visual examples (i.e., sound collages, interviews with users, short videos) as needed to complement their writing and to illustrate their focus and personal stake in the project. This assignment asks students to create a webtext with a specific audience and purpose while students develop podcasting and interviewing skills. These webtexts are based on research about sound objects or events and include interviews or conversations with people responding to those sounds.

The assignment outlines what students must include in their final project. Introductory material includes researched background information and context that goes beyond the content of the specific sound(s) analyzed but still relates to the purpose and meaning of the specific event or object profiled. Research, background texts, models, details about music, and other audio, as necessary, are required and all take this project beyond a simple summary or review. Furthermore, sonic events and objects must be uploaded, profiled, and analyzed using the reflection questions and reduced listening techniques (i.e., the sequence outlined above) that students practiced throughout the semester as a model.

Like the sonic memoir assignment, the multimodal blog project requires a microphone to record interviews. Students also include other material, such as music or sound effects, that fit their project and topic. Found media (not captured or created themselves) must be shareable under a Creative

Commons license or under fair use guidelines. When recording and producing their audio, students can use their own voices, but they contrast that performance of their voice with the other multiple voices of their interviewees and media clips, along with sound effects, background noise, musical interludes, and/or a musical soundtrack. They end by reflecting upon the rhetorical function of each sonic element and how it helps to create the theme, argument, mood, pacing, and other features of the overall project. Projects varied in genre from art exhibits to personal narratives to public service announcements (PSAs).

Combining the multiple narratives and deictic categories, Dr. Matthew Sansbury, one of Mary's former students, produced a blog featuring several sound collages and reactions to the "It Gets Better" movement, including PSAs airing on radio and television against the bullying of LGBTQ youth. For his senior seminar capstone project, Matthew entitled his project "It Better Get Better" to stress the rhetorical valence of his message and his activist deployment of this phrase based on responses to his public service messages. Using the affordances of social media to pull followers from his other digital presences to this website, Matthew gathered initial responses and coming out stories to incorporate into the blog as a model for a living multi-valiant document with powerful exigency. Matthew's written reflections on this project, excerpted here, included many insights on the exchange practices that occurred throughout:

> *It Gets Better* provides a chance to research the interaction between writer and audience, and how the audience forces a writer to [rethink and even] restart the process . . . I wanted to utilize the concepts and theories of traditional, visual, and sonic rhetoric to analyze the *It Gets Better* movement as a local exigency and represent the phenomenon to my target audience: university students and faculty . . . I decided to include a podcast short on my web text in order to give the weblog more variety . . . As an ethos-builder, I chose to design the web text to be an edgy expression of my aggravation with the social climate of our times. For the logo (and logos), I chose easily recognized symbols and edited them together with Photoshop: earth, heart, and peace sign. Ultimately, in all the pieces that constitute the site, I chose elements that would make my audience feel comfortable and supported at the same time.

Matthew's thoughtful reflections describe in detail his many successful choices about sonic rhetorical exchange practices and audience research and digital production. His self-assessment is confirmed by his own audience responses. As teachers, we base our assessments of sonic rhetorical practice and compositions upon the students' exchanges with their classmates and other audiences, as well as their individually written, spoken, and recorded reflective statements. Keeping a web site active, especially as a living document tied to an activist exigency, is an ongoing challenge that leaves projects like this one and its intentions behind to represent a moment in time. Matthew's successes with sonic production, visual arrangements, digital design, and crowdsourcing audience responses, as well as his reflections on learning, all helped us to confirm the educational value of these sequenced sonic rhetoric assignments.

Resonant Listening as a Transformational Cultural Practice

Through classroom and professional websites, students always exchanged their audio documentaries and multimodal blogs with others, thus locating themselves not only within local and regional soundscapes but also within larger networks of composers and producers. Traditional soundscape documentaries, like Westerkamp's Kits "Beach," are meant to be consumed as fixed media—a fairly static representation of the Vancouver beach in 1989—during a pivotal moment in the city's development. In contrast, the soundscapes produced by our students became reframed as part of a larger regional collaborative mapping project. Individual clips and remixes were put in direct conversation with each other to create a diverse, participatory soundscape archive. Such an exchange helped students locate themselves as both composers and active participants within their regional soundscapes.

Sound studies, reduced listening, sound art, music, podcasting, and sonic composition experiences have all provided us with the forms and language to guide students toward richer perceptions and experiences of sound itself, as well as our many cultural ways of listening. This resonant listening approach can make them more receptive to the effects of not just linguistic content but also tone, pitch, and vibration and make them more attentive to resonance and dissonance within their various soundscapes. Inviting students to postpone their (trained) analytical responses and focus on the immediate material aspects of sound objects and interactions is the first key step to experiencing and understanding resonance in these exercises. The practice of postponing also brings those disciplinary and cultural ways of listening into relief, making them available for reflection and transformation. Finally, the approach invites students to understand our soundscapes as cultural processes for producing and listening to sounds. Situating them as cultural makes them even more susceptible to disruption and change.

An important outcome of the sonic pedagogy sequence is that students produce, exchange, and distribute collaborative soundscapes and sonic archives and, thus, witness their own sonic cultural engagement and transformation. At the very least, we hope students leave this experience-based sequence with an understanding of sonic arts, voice-overs, and podcasts, not simply as technical or narrative devices that stand over and against cultures and landscapes, but also as a means and a metaphor for composing many genres and experimental texts in general. We would like to thank our former students Matthew Sansbury, Sonja Engelstad, and Kelsie McWilliams for their generous classroom and research contributions and their willingness to publish these sonic projects and processes as public-facing websites. In the next two chapters we describe and analyze how our collaborative work with musicians and artists has helped us develop and refine this sonic pedagogy by expanding listening practices to include forms that make us even more receptive and sensitive to the human and nonhuman agents vibrating within our environments.

3 Exchanging Songs in the Song Circle and Recording Studio

It's something very transcendent about allowing yourself to be a vessel in the moment for the music that's always in the air . . . It's a current that's there and you can dip into the stream. We do it with our dreams, with our thoughts, but to do it with music, it's terrifying but transcendent . . . and it can be a way to give that to the people so that when they leave they can live free and truly exist.

–Jon Batiste, *The Late Show with Stephen Colbert*

Musician and former *Late Show* bandleader Jon Batiste, congratulated by host Colbert for his recent awards and concerts, responded with this exuberant description of tapping into an existing "transcendent stream" of music that he gives "to the people" so they can "live free and truly exist." Tapping into and channeling an existing musical stream (a belief among some musicians) points to the fully embodied and powerfully emotional experience of music on nearly all human beings. While some may question Batiste's ethereal explanation or the assumption that music is a universal language, music as vibrational experience always affects embodied human experience with its evocative emotionality and its rhythmic effects on the entire physical body. Specific sound frequencies can deliberately affect mood, and more. After all, bodies resonate in nonverbal and creative ways as sympathetic harmonics move flesh and bone, ring out on instruments or in voices, and whisper familiar rhythms and melodic themes. Even when unaware of these impacts, some may psychologically experience something they might call transcendent. Such a resonant experience among musicians and their listeners, suggests Batiste, can lead to freedom and inspirational self-expression for those who convey the music.

With our own participation in sonic arts and our students working in sonic media formats, we also hope for moments of inspiration leading us to self-expression that is fully embodied and engages our emotions and empathy. This is how we define and teach resonance—sonic experiences where the multimodal bodily senses are just as important as the reasonable reactions and the expected linguistic responses in our classes. By following the assignment sequences offered in chapter two, we introduce assignments and experiences that increasingly moved away from linguistic-based audio essays and rhetorical texts to engage students in sonic processes for self-expression, cultural soundscapes, and musical experiences. These experiences are based in locations: they depend upon listener interactions, and they are subject to one's positionality in culture. Musical experiences offer us additional evidence of regular sonic practices and new pedagogies, as important collections by Michael J. Faris, Courtney S. Danforth, and Stedman (all published in varying authors' order) have elaborated (e.g., *Soundwriting Pedagogies* in 2018, *Tuning in to Soundwriting* in 2021). We expand upon new collections of sonic practices by studying additional locations and sonic subcultures. In this chapter, Mary and her musical colleagues engage in resonant listening and ritualized cultural practices of receiving, amplifying, recording, and exchanging songs. New locations and practices continue to push diverse sonic rhetorics further, and this chapter explores in detail how songwriters use resonant listening and embodied responses during musical events to form their central sonic experiences. Those practices, in turn, offer us new lessons about pedagogy for self-expression, collaboration, and audience response.

A song itself, far from just a "text," is a process grounded in time and place, yet a song also has fluidity in its recursive and unfinished nature. Halbritter explains in "Who Wrote That Song?" that songs are continually rewritten and benefit from increased possibilities offered by the many performances, audio productions, and circulations in which songs are reproduced and re-performed. In fact, this networked process emphasizes an ongoing "storytelling about lived experience" and parallels Donald Murray's concept of "unfinished writing." If, like Halbritter, we ask, "who (all) **keeps writing** that song?" [emphasis in original], that question makes the process visible. Halbritter's additional discussion of recording studios as compositional spaces challenges, among other assumptions, the dominant model of publishing musical products over the process-filled life of a song as expressive exploratory learning. Expressive language, based in foundational composition studies research, depends upon discovery and trust, as Britton explained in an early *National Writing Project Newsletter*:

> [W]hether we write or speak, expressive language is associated with a relationship of mutual trust, and is therefore a form of discourse that encourages us to take risks, to try out ideas we are not sure of, in a way we would not dare to do in, say, making a public speech. In other words, expressive language favors exploration, discovery, learning." (9)

Songwriting as an expressive practice captures these experiences of trust, risk, and the self- and audience dynamics that occur within exploratory composing. And yet, even lifelong writers and songwriters like Mary can find it hard to have the confidence, the trust, and the freedom, much less the inspiration described by musical prodigies like Batiste. Songwriting, like all writing itself, is difficult; it takes dedication, training, and practice regardless of the musical genre. A musician's rhetorical framework begins by resonating with certain sounds and songs, as well as with their own voice and prior training. For example, VanKooten (2016) offers a powerful video reflection on her embodied singing and composing in the webtext "Singer, Writer." Here, she enacts her rhetorical resonance and embodiment while singing with a choir. She opens with "connection" and moves through increasingly discordant sounds: "I sing, I write Choric, connected Embodied, Filled with emotion Together, with others I sing. I write. I compose." This spoken word performance, which includes samples, sound effects, and remixes from research presentations, recomposes the music, noises, and bodies that together illuminate how listening to and reflecting on your own voice through recordings offers a powerful self-reflexive cultural insight into singing experiences.

Inspired by Halbritter and VanKooten to reflect on musical experiences more deeply, this chapter describes Mary participating in and examining the musical applications of the receiving, amplifying, recording, and exchanging sonic practices within specific locations. These examples focus on two key locations—the song circle and the recording studio—where events with other professional acoustic singer-songwriters occur. After playing music in structured song circles for years, as well as recording and producing music in studios, Mary discovered that songwriters share common sonic practices that are based in moments of empathetic connection.

This study of songwriters focuses on their roles as expert professional musicians in public locations, and it analyzes routine events and cultural rituals with fellow musicians who are experts in writing, teaching, recording, and performing music. Mary has played music and recorded with this group of singer-songwriters for years. They are songwriters who all have been composing, playing, recording, distributing, and performing songs in various combinations with one another. Using "convenience sampling," Mary attended events with musicians who have all published original Americana/ folk music albums and with whom Mary has collaborated on musical projects (Gentles and Vilches 2). Mary explicitly sought out events that shared her focus on acoustic folk, Americana music so that while participating in these activities, she could deepen and reflect upon her own perspectives.

The recording studio looms large as an early experience because the process of recording and collaborating with other songwriters completely changed Mary's embodied experiences of singing. When Mary began recording vocals with another songwriter during graduate school in 1991, she learned how to use a microphone with noise-isolating headphones, which allowed her to hear the tonal qualities of her voice amplified for the first time. These new forms of singing, listening, and recording vocal harmonies had Mary resonating with her own voice in a completely new way. In those days, the producer layered recorded music tracks for each song and had musicians "punch in"—record short improvements to a vocal line or an instrument at that place in the linear recording—as well as manipulate recorded sounds using digital audio effects like auto-tune. The studio became an outlet for Mary's earliest songwriting colleague, a pioneering woman songwriter who had begun featuring women in her recordings while also founding a women's coffeehouse that became a major outlet for performances by women. Indicative of the times, she used her radio program hosting, her live concerts, and her cassette tapes to distribute her own and other women's songs. More importantly, the music studio helped to nourish a community of songwriters, radio program guests, and performers that included many more women than in any studio's previous recording sessions.

Mary's earliest studio recording experience showed not just the importance of a music studio's culture, but also how musicians recognize that interpersonal dynamics, including gender dynamics, play essential roles when collaborating. These dynamics can point to what works and what doesn't work in practical musical settings. The music production and airplay industries, long known to be male-dominated and unequal for women artists, have left women artists neglected or ignored in music, thus generating microcosms of unequal gender relations and representations in nearly all musical subcultures. To give one example, sociology research on women's popular music demonstrates how women fail to reach the status of "cultural consecration"—Bordieau's term for cultural acts that venerate the chosen few who can be creators within forces "shaped by existing cultural frameworks about art and gender" (Schmutz and Faupel 685). The "unconsecrated" push against this cultural framework and recursive process of an oppressive musical culture.

To fight against the persistent gender inequality of male-dominated musical communities and industries and to help usher in more feminist possibilities for these sonic practices, Mary sought songwriter colleagues who self-identify as female and/or queer. In this group, four are cis-female, four are cis-male and one identifies as a transgendered man. Five songwriters identify as LGBTQ+, while everyone identifies as white. All are middle-aged or older, ranging from ages forty-two to seventy-eight. Even within this rather homogenous group, naming different forms of participation and honoring multiple ways of experiencing music helps to reflect and amplify each person's subcultures and intersecting identities.

The Song Circle as Cultural Activity

As an informal gathering without an official leader, song circles do not operate as sing-alongs or open music jams, but instead emphasize the equal standing of all songwriter colleagues and the committed opportunity for each person to take their turn. While a jam presumes everyone will play along together and join in as best as they can, song circles have a basic formal structure that assumes equality and acceptance among participants. Singer-songwriters have an opportunity to explore their identities and forms of collaboration when in a structured group like a song circle. The song circle has basic rules for operation and always aims to provide a safe and supportive space for playing a song and perhaps receiving musical feedback or other notes on the writing. Or group members can decide they simply want to trade songs back and forth with no overt response or commentary. In the context of an individual songwriters' circle, members will establish the ground rules for playing (e.g., "no excuses") and receiving feedback (e.g., "start with something positive") when a writer offers a new song for feedback. Songwriters in the song circle will share the musical key of the song for those who may want to accompany or play an instrumental solo. The individual musician decides what kind of feedback they want to receive, whether a response is required or even discouraged, and whether improvised accompaniment is welcomed.

The enculturation that occurs in a small group setting depends upon shared experiences, language use, deep listening, and empathic understanding, alongside a set of guidelines for playing together in an equal and organized way. The specific locations of song circles and studios are places where informal social music is ritualized within the everyday folk cultures that have sponsored such events and, as such, have their own culturally defined contexts and signal sounds: in the bar that hosts an Irish session, at the coffeehouse that offers a blues jam, and within the spaces of festivals and small music venues that feature performers.

These regular rituals help define the group's cultural features, and for a singer-songwriter's group, culture emerges in the routines practiced and the rules followed. The rules established for a song circle are based on experiences with folk music cultures:

- First, like a public "open mic" performance at a coffeehouse, different musical abilities and styles are welcomed so everyone gets a turn while others must listen.
- Next, any accompaniment must be requested, or if allowed to occur spontaneously, must be added carefully so as not to overwhelm the singer or their song.
- Third, no one person dominates the song circle as expert or leader, so any facilitator is simply another songwriter.
- Finally, while songwriters may request specific feedback up front, others simply need to work through nerves and get through their song. Groups can use guiding instructions like "no excuses or introductions needed; just play the song."

Everyone's goal for trading a song is important but does not have to be identical. These general rules, which are not exhaustive, allow for adaptations and detailed instructions, and they offer the chance for a group of songwriters on location to build patience, confidence, comradery, and most importantly, trust.

In coffeehouses, parks, living rooms, and front porches; a group of singer-songwriters will gather, often on a weekend afternoon, to exchange songs. One such song circle took place at Mary's house on an October Sunday afternoon in 2019 (Fig. 1). With food, drink, and dogs included, the casual gathering established comfort and ease for participants. After noshing on treats and opening some beverages, the group established common goals and individual preferences: they would go around the circle twice so each person could play two songs uninterrupted. Resonant listening that a photo cannot show occurred through their listening ears, vibrating bones, swaying heads and bodies, tapping feet, guitar-picking fingers, and singing voices as they responded to each song. Playing together generated a resonance created in that specific moment and enhanced by the slightly structured ease of the location. That day, the song circle included the possibility of learning individual songs, since a teacher and songwriter had brought one or two songs with print lyrics to share with the group. Huddled over lyrics sheets, the musicians all took notes to play or to sing along more easily to that person's original composition (see Fig. 1).

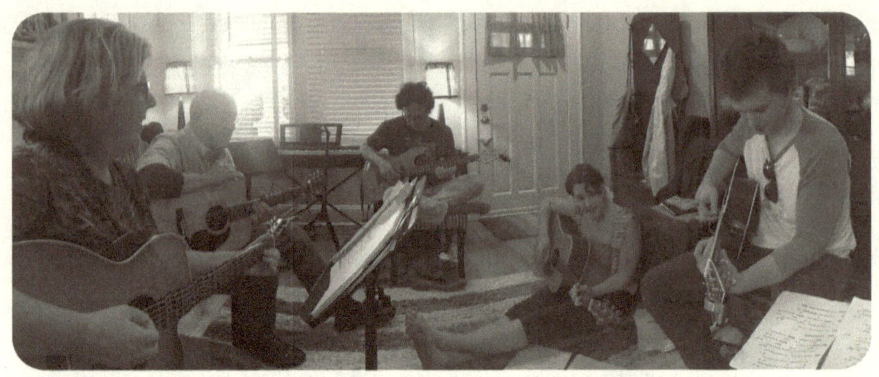

Figure 1: A Song Circle in Mary's Living Room. Photo by the author.

The other singer-songwriters, as receivers of the song, both amplified and reflected back the lyrics. In acoustical science terms, the amplification and reflection of sound waves are almost always directly related. A sound reflects or bounces off surrounding surfaces in the space to varying degrees (hard surfaces creating the most echo and fabrics absorbing the most sounds as experienced in any restaurant). When singers put on a harmony, they've amplified and "thickened" the sung melody through harmonic contrast, and they've also made the combined singing louder. Amplification also relates to the structure of song parts and their contrasting sounds. Acoustic folk and Americana music emphasize storytelling, with the words woven among traditional rhymes that are typically divided into verses and a repeated chorus, often with a one-time bridge and instrumental solo but also with many variations in structural possibilities. Playing a solo with a guitar or violin as an instrumental break means other instruments and the singer-songwriter must listen and hold back the volume; they play backup chords or rhythms during the solo. This process further amplifies the contrasting parts of a song (like the sung verse or the bridge) and adds textures that can contrast the song's other instruments. The dynamics of this exchange are complex: the musicians, while improvising, must make eye contact and listen, back off when necessary, and take their cues from the singer-songwriter who is playing the song.

This group of colleagues included several advanced instrumentalists, singers, and music teachers, so those people quickly offered to play or sing harmonies along with others. They began with an original song that the writer, one of Mary's producers, calls his disco song, from his newest album. Everyone listened in awe as he flawlessly played difficult chords and sang in an unusual falsetto voice. Moving around the room, each person played one song, with some of the more experienced guitarists and singers joining in from time to time. Another original song was an anthemic tune that everyone learned together on the spot, with lyrics balanced in front of each player. As a result, the song quickly included a chorus of voices and instrumental accompaniments, and the songwriter led the way, signaling an extra two instrumental verses for individual guitar solos. This songwriter and music teacher is accustomed to conducting ensembles and organizing group performances, so she ushered in additional parts with ease and together, the group amplified the song. Afterwards, she exuded confidence about her musical arrangements.

Not everyone in this group, however, has that confidence for improvisation or the vocabulary for creating musical arrangements on the spot. The complex and sometimes uneven dynamics among individuals in musical improvisations can be intimidating or silencing to some musicians. Two of the songwriters were intermediate-level guitar players and had limited experience playing with other people, having just learned as older adults how to listen to and play along with other instruments. Still others were reluctant to sing for fear of forgetting lyrics or missing guitar chords. Not being intimidated by more advanced musicians and grappling with the nerves of performance anxiety are important goals for many songwriters. As each person played, some joined in tentatively and some simply listened. Playing the music together puts the song into dynamic circulation, where it is rewritten and reverberates into many forms.

Sonic techniques of songwriters include the specific moments of improvising or collaborating through resonant listening, receiving, amplification, recording and exchange that, when taken all together, become a spontaneous response to the song. Songwriters recognize that the fluidity of moments when playing music together can push a song into a new form. Each moment of working with a song includes cultural forms of exchange not just with other songwriters, but with a wider cross-section of the songwriting and folk music community. Most importantly, the music producer-engineers are indeed experts, but songwriters must step up and be experts in order to collaborate on recording their own music. These moments highlight the experiences of trust and the risks that can occur for songwriters, but song circles offer the best generative and supportive practical environment for tentative experimentation.

The recording and performing that ultimately occurs with a published song involves the public elements of musical popular culture and may launch the song into an ongoing rhetorical circulation. By way of resonant listening practices generated from listening and exchange in a song circle event, a community of people will evoke a specific cultural soundscape through the ritualistic collaborative practices for receiving, amplifying, (re)composing and exchanging each song. When moving from an intimate circle into the recording studio, the many tools (beyond instruments and voice) for capturing and distributing songs and the new location usher forth even more sonic practices.

The Music Studio and Reduced Listening

The two audio engineers and music producers Mary worked with on her original music began their careers early by recording in their family's basements as teenagers. Both now have well-equipped professional studios and a devoted following: they make much of their living producing and engineering music for others, while adding their own musicianship to others' recordings since both are accomplished multi-instrumentalists. Their work involves careful listening; identifying tones, rhythms, instruments (including voices) and styles; and internalizing the "parts" in any composition. As highly sought-after producers, they both offer support to musicians, and they perform locally, regionally, and abroad. Mary's first producer, trained in eastern Indian and Turkish musical styles, tours internationally and plays world jazz music. He was able to engineer, help arrange, and produce Mary's album, *Spinning the Sky*, by playing multiple instruments and recruiting local session musicians. The other producer has worked with four of the other songwriters: he produced most of the musical arrangements. These producer-singer-songwriters and multi-instrumentalists overlapped as experts in both locations—the recording studio and the song circle—because of their various musical relationships among the songwriter colleagues. As producers and engineers, these two artists filled others' recordings with instrumentation from nearly a dozen instruments before completing the post-production mixing and mastering of individual albums.

Though the songwriters' styles vary widely, their musical cultures overlap within various Americana and acoustic traditions, ranging from traditional Irish ballads to New Orleans street music. Mary composes guitar-based lyrical songs that draw on jazz and blues vocal styles that emphasize her vocal range. One producer composes and plays a wide range of Americana and folk pop music, while another producer performs solo acoustic folk guitar and also world jazz music in an ensemble. One songwriter writes traditional American folk and country songs and has recorded dozens of basic solo tracks (inspired by heroes John Prine and Kris Kristofferson) to document his personal repertoire for posterity. Yet another songwriter composes acoustic and electric blues songs inspired by New Orleans funk and roots music. The teachers and producers of the group all have advanced training in jazz and classical music, and each has performed in a variety of rock, jazz, and pop bands, in addition to their more lyrical acoustic songs.

Each producer has a home-based studio of dedicated rooms separated by multiple panes of glass and with walls covered in tiles for sound isolation. A soundproof glass booth isolates the artist to record new tracks while wearing headphones that play previously recorded tracks as a guide. The studio walls are crammed with instruments and cables hung on hooks. The producer sits in a wide corner that is filled floor to ceiling with electronic equipment and at a desk with a computer running the audio editing software. He's surrounded by a microphone, keyboards, and electronic effects. To work, he carefully listens while he watches the visual representations of recordings as sound waves in audio editing software (see Figure 2).

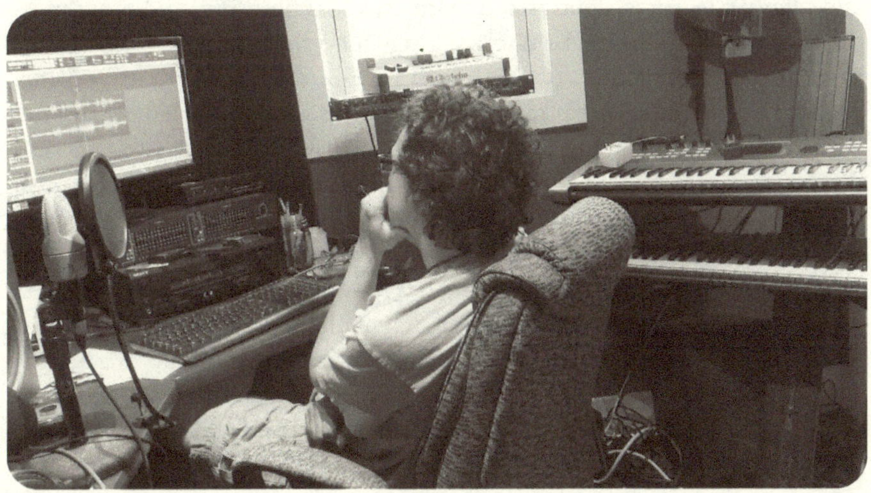

Figure 2: The Studio Producer Listens to Mary's Song While Watching the Soundwave Tracks. Photo by the author.

The producer's receiving, amplification and recording of song elements begin at the desk. First, he sets the appropriate tempo for the song and creates a click track—a steady beat generated digitally that supplies the rhythm within which all the music will be recorded. This click track structure can be a struggle for an instrumentalist without much recording experience or who, like Mary, plays in a looser style without a band or any percussion to ground their rhythms. Sounding surprisingly like a compositionist, but also like an engineer, the producer emphasizes how the initial click track provides the "scaffolding" for the entire song. Mary and the producer begin by creating an initial reference recording at speed, called a "scratch track" for the song.

The recording session began when the producer set up a matched pair of condenser mics and carefully placed them near the sound hole of the acoustic guitar, reminding Mary to take note of the placement when recording. The microphones receive what they are set up physically to "hear," so mic selection and placement is, once again, essential. Singing separately in a sound-proof booth with headphones to a guitar track can be disconcerting at first. Refinement of the song sometimes occurs during recording but usually at a new session, like one they had later in June for adding vocals and mixing the recorded versions together. In the mixing process, the producer captured three sung versions of a vocal line, then listened with Mary, line by line, for the "best" version and saved each section onto a new track. Listening together, through a form of reduced listening, allowed them to isolate and identify textures, contrasts, dynamics, vocal phrasing, and other musical qualities. Careful music track editing smooths out the cuts and transitions among recorded parts. When remixing music, songwriters and their producers work extensively to amplify certain sounds and shape them into a desired arrangement and delivery of the song. Given the flexible physical-digital possibilities in contemporary digital recording, producers and musicians can engage in multiple sessions of this sonic layering while they record and remix their music at home. As another songwriter recording in the same studio explained it, songwriting has lessons about the synergy between words and the musical arrangement that will support those lyrics. Being able to work with arrangements at home electronically and do pre-production work on a song, however, makes the process much better in the studio.

The shy lifelong songwriter's experiences in the studio demonstrate a different set of physical and psychological challenges for a self-described analog person. As someone who has written songs and played the guitar for over forty years, he never played his songs for anyone until he began music lessons and attended a monthly gathering at one of the music teacher's events, where he also met other songwriters for the first time. Inspired by the musicians' comradery, and the oldest songwriter in the group, he began recording his first album of original songs at age sixty-nine. Collaborating with the producer, together they created newly arranged song versions with guitars, bass, and percussion, while other instrumentalists were hired to contribute expert violin and harmonica parts. Even as they stayed close to the song's original structure and lyrics on the printed page, they also made modifications based on the studio experiences and feedback (see Fig. 3). Through this process, producer and songwriter together decide upon the final structure, tempo, and rhythm for a song. With each new instrument part added, the song evolves into a new shape. The songwriter grew familiar with what the sound of a mandolin, a harmonica, or a pedal steel guitar might add to each new recording; and through conversations he and his producer shaped each song, session by session, into an arrangement of instruments and voices that convey the song's story and feeling. Mary's experience with both of her producers was similar in that she had little experience with creating new instrumental arrangements and restructuring her own songs. Mary, like the other songwriters, learned to depend upon exchanging a song with her producers and letting go of its original shape and speed, which meant giving up some amount of control over the song's recorded performance.

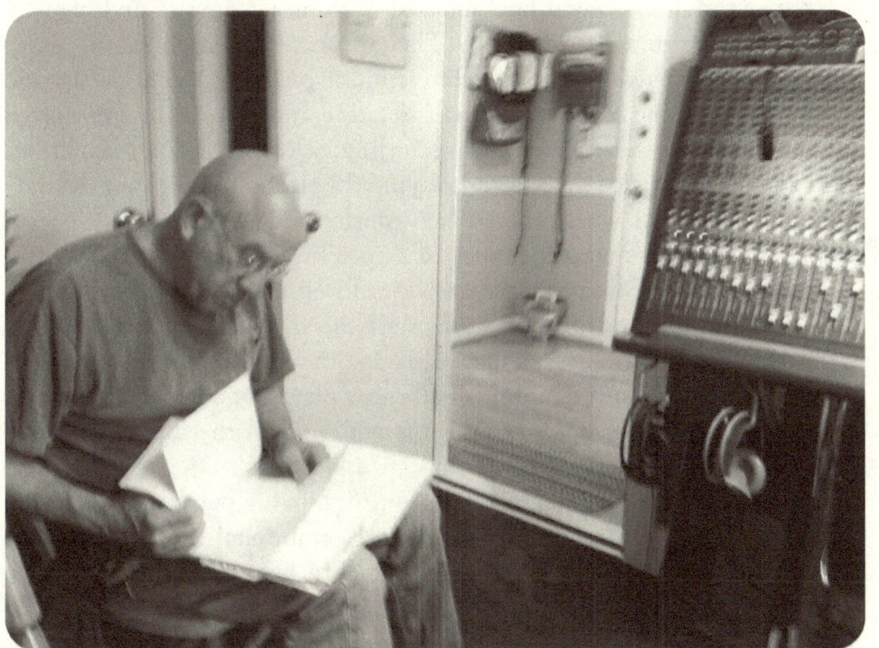

Figure 3: A Lifelong Songwriter Studies Lyrics During His First Studio Recording Experience. Photo by the author.

The resonant listening that producers use while capturing and mixing audio tracks moves regularly between reduced listening for qualities of sound and ecological listening for the reverberation of contexts. For music producers, mixing music requires focusing deeply on specific elements and parts of musical sounds in isolation: the tones of a human voice, the timbre or unique textures of each instrument's voice, and the delivery of each moment of a performance. The experienced songwriter with his producer and/or sound engineer then re-contextualizes each individual sonic element back into the composition, while working within an ecology of resonant listening. This kind of "listening like a producer" takes training, practice, and draws upon cultural conventions—the existing songs that can provide examples of what sounds like the desired result. This intense focus required by reduced listening guides a musical composer's perceptions as she isolates the tonal, rhythmic, and amplified elements when mixing together the recorded tracks. The process requires a parallel stepping back when listening to the whole recording. The felt quality of one's awareness reverberates from the recording and its projection or amplification through speakers, and then the sounds bounce back or otherwise return to the body, marked by sensations and recognitions.

Mary's producers reveal how musical training is complex and helps to define one's cultural background. Their expertise exemplifies an adage among recording artists that an effective music producer listens deeply and then encourages the best possible interpretation and performance of a given song and captures that moment in a recording. Studio recordings usually require intensive practice singing and playing the elements until they "sound" right—the moment where the trained musician and experienced listener resonate with their own experience of a specific musical culture and cultural soundscape. One producer describes listening obsessively to records while growing up and breaking down all the instrument parts. He also credits playing in marching band and playing the French horn in an orchestra because "sitting in the middle of all those players" and hearing "all that sound" forced him to perceive and interact with all the different instruments in space, all from a specific embodied location.

In addition, the remixing process requires moving between specific details and the big picture in order to create an emotional experience for the listener. Mixing requires lots of silent breaks to hear something again "as if for the first time." This compositional process depends on silences, and upon some amount of nostalgia and memory. When describing audial experience of an acoustic sound, one producer explains how sound is always "fleeting" through its own decay, so that experiencing the physical qualities of an acoustic sound only lasts for "about fifteen seconds." Furthermore, recreating that immediate, ephemeral sonic experience is difficult, if not impossible. His observations demonstrate that what one remembers is an impression or association where sound textures and qualities are linked to an experience, a place, or a memory. When Mary asked her producer about the body's responses to sounds, he immediately laughed and said that sound is all vibration. Because we're all vibrating also, we are interacting with all of it. He mentioned how instruments like gongs, Tibetan bowls, chimes, and frame drums that have one strong pitch will resonate the longest and get people resonating, hence these instruments' role in applied sound therapies and meditation.

Several years later, when Mary went back into the music studio to record several new songs with the second producer, she had a simple premise much like VanKooten's (2016) initial hypothesis about choral singing, where "singing with others, trying to blend, learning lyrics and notes, rehearsing, memorizing, performing—all of this seemed, at times, a lot like writing." By comparing the process of blending, mixing, and performing in music studios to writing processes; Mary thought she would find similar parallels and, ultimately, lessons for students engaged in multimodal composition—or at least find a close parallel between effective teaching processes and songwriters' sonic practices.

For example, one songwriter who is also a middle school teacher of digital arts gives students the tools and workflow benchmarks, thus offering the freedom of storytelling to inspire her students' projects year after year. One could argue that an effective writing teacher does similar things—inspires students by finding ways to capture the best processes leading to the performances of students' own stories and reflective learning. What Mary found, however, was that the studio workflow more resembles a publications production process, with the producer acting as a default editor-in-chief and with a finished product's distribution as the primary goal.

Public-facing writing and professional writing projects share this goal, so a music studio could certainly offer lessons for digital production and current practices for nonlinear editing (organizing), arrangement (layout), and digital distribution (publication) of musical recordings. Many songwriters defer to the producer's expertise, however, and perhaps feel less ownership over the final publication than students feel about their school compositions. In fact, unless one is wealthy or owns studio space for limitless invention and tinkering, songwriters who pay between fifty to many hundreds of dollars per hour for recording are pressed by time and their budgets. Thus, most contemporary music production and studio recording sessions are designed to capture well-practiced and rehearsed content, rather than to offer open, unlimited space for improvising or composing together in the studio. Given the financial constraints for most working songwriters and music teachers, desktop digital audio production software and hardware inputs have afforded the most accessible means for home recording and self-distribution. As of now, even more affordable tools allow songwriters to sequence their own musical arrangements and pre-produce their recordings with a particular style of instrumentation. The goal in the studio, then, pushes songwriters toward being fully prepared and rehearsed in order to capture and mix the best possible performances in a time-driven context. This goal can also result in tensions and failures, including a recording taking too long to finish and the risk of losing touch with the original song too quickly when publication and distribution become the driving model for studio recording.

In the studio, to receive a song can simply be to experience it through multimodal embodiment and reduced listening for the first time. Responding to the song is a multimodal embodied experience that ranges from ear-based listening to body movement, from dancing to foot tapping, and from swaying to playing along with the song by using eyes, ears, software, and previous experiences with musical instruments (including voice). While artists sometimes describe receiving songs and compositions as if channeling it from elsewhere, experienced singer-songwriters emphasize how that receiving is, in fact, enabled by the hard work of practice and of writing, including writing badly. Emphasizing the importance of regular experimentation, Mary's second producer emphasized how one must practice hard and be open to collaborative transformation. Songs are shaped by a musical artist's tastes, training, and versatility, and they also are rewritten and transformed over time by an artist's willingness to take risks, experiment, and collaborate on the song.

While the singer-songwriters engage in sonic rhetorical exchange with practices intended to equally benefit all songwriter colleagues who are enculturated into the song circle's goals, they still encounter the perennial human issues of power, privilege, and access. These power relations will affect who gets to "speak up" and who is silenced. Specifically, who has access to an instrument, when can they practice, who can afford studio time, who can show up at musical events, and what support do people need to continue songwriting? More broadly, what cultural frameworks exist to support the singer-songwriter-performer and how can social dynamics be navigated to create a fair and diverse playing field for musicians?

Cultural Exchange among Songwriters

The embodied resonant practices of singer-songwriters are predicated on the tangible relationships among community building, exchange, and collaboration. Mary's findings demonstrate that creating a trusting collaborative community, establishing a process for playing music that offers equal standing, negotiating complex interpersonal dynamics, and building a trusting collaboration with a producer are all vital processes for an effective sonic composing experience. Several broad cultural arguments emerged during Mary's exchanges in the songwriting process:

- Dominant cultural norms of ability operate among songwriters, including the downplaying of inevitable and ongoing physical effects on aging musicians.
- Community building for songwriters is both momentary and is also deliberately sustained when music teachers and other songwriter event facilitators offer collaborative opportunities.
- Collaborative composing depends largely upon community but hinges upon the songwriter's confidence and the dominance of expertise among the producers, teachers, and songwriting colleagues.

Dominant Cultural Norms of Ability

All of these songwriters have played music for decades and the ongoing ability to play instruments or sing is expected until something changes. As a musician ages, her assumed able-bodied nature is, in reality, a culturally defined and temporary state that is always changing. Well-known musicians often lose their voices or their dexterity for instrument playing and performance, sometimes in profound ways. Famous musicians have famously adapted their voice (Joni Mitchell) or re-learned an instrument (Lucinda Williams' guitar) after a stroke, while others gave up music altogether due to physical issues (Julie Andrews). Likewise, these songwriters all talk often of the impact of age and health on their ability as they engage in their regular practices of collaboration and exchange. They bemoan their lost vocal range, neuropathic pain in arms, hearing loss, difficulty standing, inability to lift equipment, and other impairments that no amount of musical practice can correct. Even more so, adjacent difficulties with anxiety and depression affect their mood and confidence in their music.

By way of example and by following Ceraso's "Sonic Scenes of Writing," Mary's embodied experience as an aging musician is illustrated in the following sonic scene:

> *It's early Saturday morning in my living room and outside, the neighborhood is quiet except for the usual twittering and cheeping of bird chatter. The multi-tone ringing accompanies me, as it always does, to my chair. No arm pain—now that is a good surprise. I pick up the guitar and strum the strings; strong chords ring out and spark unspoken feelings. I lower the bottom E string to an open D, adding a drone that thickens and sustains the open D and G chords. The harmonics surround me, drowning all sound out except my voice as I begin to sing loudly, "Sweet Melissa"! Then much more softly, when I hear and feel the chronic hoarseness in my throat. This is not about the singing. The*

guitar vibrates my chest while I play through the song and whisper its lyrics, the empty sound hole not filled with a microphone, but its strong wood resounding, crafted from longtime care and expert workmanship. My quivering heart opens up and a trust emerges, a sense of my expressive self, and finally, some self-compassion and acceptance.

As the song fades away, the high-pitched ringing returns, tones ringing out seemingly on each side of my "ears" but really, they are inside my head. Two fans set on high fade back into my soundscape, their comforting hum blending into and masking the ringing.

To compensate for my tinnitus, I drown myself in resonant sound, then I put down the guitar and walk away. For me, listening requires a constant attentiveness to different spaces and a present mindfulness. It also increasingly requires subtitles.

Ceraso's *Sounding Composition* introduces three sonic scenes as evidence of her interactions with sound in everyday life and offers the crucial insight that "sonic experiences engage much more than our ears and brains; they also affect our physical and emotional states" (2).

Background noise and signal sounds are easily scrambled for someone with tinnitus, and this leads to frustration and avoidance of many indoor spaces. When the human voices are at a relatively similar volume and particularly if the space reflects the sounds off multiple hard surfaces, the soundscape becomes an indistinguishable mashup of noise and a source of anxiety for someone with tinnitus. In "Sonic Scenes of Writing," Ceraso describes a similar experience for one faculty member she interviewed: "For a faculty writer who recently developed chronic tinnitus in both ears . . . [b]lending internal noise with external noise allows the writer to be less distracted by tinnitus so he can focus more intensely on his writing. This notion of blending together sounds to create a non-intrusive sonic environment appeared throughout my interviews with faculty writers" (318). Ceraso later concludes that "sound seemed to carry more significance—and have more consequences—for writers with auditory impairments. Writers with auditory impairments described sound as sometimes being physically uncomfortable in addition to being distracting; sound seemed to play a larger role in these writers' bodily well-being compared to hearing writers" (330). This is also the case for singer-songwriters and to compensate for tinnitus, musicians drown the noise out with an amplified or mixed soundscape. For the instrumentalists with neuropathy, physical therapy and nerve treatment can offer some relief and possible retraining of instrument techniques. Emotional health and psychological well-being can be more elusive, especially since musicians often turn to their music for precisely that support. Research on music and singing demonstrates how both personal and community engagements with music can have many health benefits (Stacy and Sublet). However, the difficulty for singers and instrumentalists to maintain their physical and vocal ability cannot be overstated; it inevitably becomes more pronounced with age.

Community Building and Collaborative Composing for Songwriters

An accomplished and versatile musician who has taught guitar professionally since 1992, another music teacher in the group exudes expertise; yet, he also struggles with aging issues, as well as the painful experiences of coming out as trans in a conservative southern culture. His musical groups have included several queer-identified jazz ensembles and rock bands, but he is an accomplished acoustic singer-songwriter. Previously well-known as a female guitarist, he now says, "I'm the boy with the voice of an angel," referring to his first soprano voice, which contrasts with his trans-male identity (see Fig. 4).

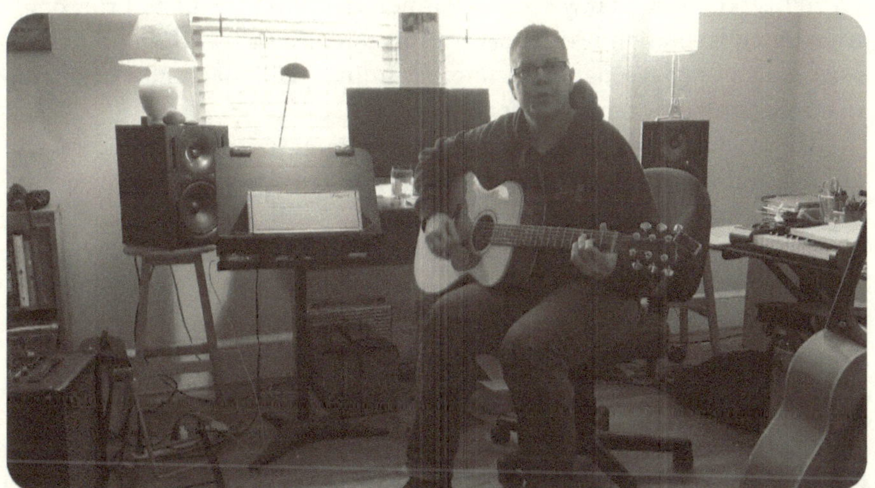

Figure 4: Longtime Guitar Teacher Demonstrates a Chord Progression During a Community Event. Photo by the author.

As the most experienced music teacher in this group, he has made community-building central to his work. A monthly "guitar pull"—his term for a Sunday afternoon song circle at his house—features a sing-along, then individuals are each invited to play one song, plus a second, if time permits. In his song circle events with his students, he deliberately builds community by having the group play "Wagon Wheel" together, a popularized folk song, to help establish trust at the gathering. Freeing up the group to play and interact with a familiar song can help build even more trust and freedom among the songwriter colleagues. Each person then has an option to play for the group and receive feedback. The eldest songwriter in this group attended the guitar pull years ago and met Mary; he also played an original song "The Table" for the group. He received much praise but humbly stated, "I'm just a thumb strummer." This songwriter took up guitar decades ago because he just wanted to play songs for himself. He also builds community but in a modest and private way. His kitchen table is his chosen place for listening to music (and "often dispensing wisdom," he likes to add). He even wrote his song "The Table" to capture its importance in his musical and social life. Mary met at his table with him many times to review and discuss his songs in each set of newly recorded tracks from the studio. As the shyest of the group, he downplayed his abilities. But, after recording in the studio for over six years, he can now recognize which musical elements he wants to include or exclude based on close listening to each new mix of the song. Together, he and Mary could build a vocabulary together to talk about arrangements and desirable instrumentation. In other words, an understanding and a trust has been built through experience where he can now say to his producer: "How about a harmonica on the opening of this one?" Another collaboration was built with this song over time, through the uneven processes of resonant listening, recording, remixing, and exchanging that song with other songwriters.

All the musicians in this group share enough common musical background and cultural experiences with music that they can resonate with songs they've not played before. Through close listening, each songwriter can still recompose a song into yet another version and blend their cultural expertise. Two songwriters who live across the country from one another, for example, can meet up occasionally and play together in a song circle. In some cases, the decades-long experiences of listening to one another carefully can quickly bring the musicians together (see Fig. 5).

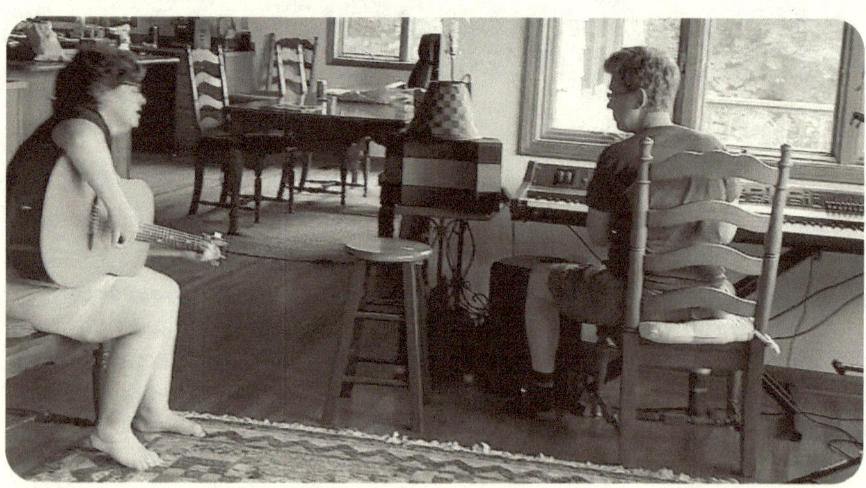

Figure 5: Opposite Coast–Dwelling Songwriters Play a New Song Together During a Songcircle. Photo by the author.

Their resonant listening included feet tapping in time to the song's rhythm, eye contact, and their bodies leaning in toward each other. Within minutes, both were singing the vocal line, with one playing expertly on guitar while the other played chords on the piano, enacting a receiving and amplification through the sharing of this song.

As one of those two songwriters later explained it, these moments of playing music together are always "relational": the events are relationship-based and depend upon trust and agreed-upon norms among songwriter colleagues. The teachers in the group tend to lead the way for building musical community. Another longtime music teacher describes "trust and support" as fundamental to her students' experiences at the performance "salons" that she carefully orchestrates. This teacher determines the songs to be learned and played by beginners and experienced musicians alike; she helpfully accompanies them on guitar or piano and with her voice. Encouraged and supported by these kinds of connections that inspire trust, songwriters can gradually learn to perform on their own and engage their audiences. The teachers within the songwriters group all believe they must play a central role in building and maintaining a supportive community for their own students. Their guitar students and players range from complete beginners, who need basic practice playing for others, to singer-songwriters who need feedback and expert help on their songwriting. The students and collaborators who write songs and attend these events gain trust and a willingness to take risks when supported by these community-building sonic activities.

Music is used in diverse cultures to celebrate and to soothe, to excite and agitate and to testify. Singing and songs are embedded into many highly significant cultural experiences and generate moments of connection within multiple locations and contexts. Music evokes responses and is also central in many cultural events where the responses are more formally ritualized and controlled, like bat mitzvahs, graduations, speeches, and funerals. At the same time, intensely personal experiences and emotions can be evoked by music: the passion in singing along to a beloved lyric, the joy in dancing to a favorite song, the fists pumping in response to a live concert, the calm found in a chanting meditation, and the pleasure in playing a well-known instrument from muscle memory. These experiences with music are often described, similar to athletics, as being in the "flow," and a flow that is not just fully immersing oneself, but also focusing attention and letting go of the thoughts. When attention comes into focus, that moment opens up a resonance with oneself and that accompanying freedom to let go. Attentiveness is, in fact, a most effective tool for moving through pain, through conditions like tinnitus, and also letting go of the pain and tuning out the noises.

Lessons about songwriters' coping mechanisms can also translate to the complex decisions and interactions of communication. Musical improvisation, for example, allows a musician to connect in moments of attentive, spontaneous collaboration that is also marked by a kind of "flow" and letting go. Improvised musical exchange acts as both a response and a remix that can reshape a song and make previously unheard musicians feel individually "heard" rather than silenced. In one example, Burke scholar Gregory Clark invokes jazz musicians as a prime example of American democracy, marked by the spontaneous interchanges that happen when players enter the ongoing conversation in the imagined space of the Burkean parlor and engage in identification. Clark considers this image as the ultimate American democratic experience of individual expression balanced carefully with working together in collaboration "toward a common goal" (19). These spontaneous moments of improvisational engagement, then, can help sustain the songwriters' individuality and build their community in those moments without calcifying them or creating absolute borders like musical genres might suggest. Musical moments will shift into circulation in a variety of ways and still carry the traces and reverberations of each moment and each location.

The song composing process happens everywhere that songwriters go as they operate within engaged communities that form, dissipate, and then re-form through musical events and the exchanging of their songs. Song exchange practices can be ephemeral because exchanging the song exists in the exchange itself and in the situated moments that musicians come together to play, record, or perform. The rhetorical lessons we can draw from these sonic practices help add to our current understanding of situated, everyday composition. Songwriters' everyday practices around their songs makes them a community that forms around specific moments and interchanges, which are impacted by cultural conventions and community expectations. The conversations and participation in the song circle and the studio help to identify sonic practices, as well as lessons about the types of community engagement taking place within the day-to-day sonic activities of the songwriters.

In the cultural space of song circles and recording studios, the pre-linguistic conditions for rhetoric and the role of the nonhuman in rhetoric operate alongside the physical and digital texts and performances that are taken up in an ecology of circulation. Hawk provides an apt example of this when he examines the alt-musical group Refused's album *The Shape of Punk to Come*, and demonstrates how the recording gets caught up in its circulation and musical "worlding" and in the process, creates its own complex "sphere publics" that underlie its eventual success (199). Musical experiences, like those of Refused, can offer agency and acquired cultural resonance when fostered and nurtured by deep immersion in sonic practices and then released into ongoing networks of rhetorical circulation. Singer-songwriters likewise do the work to create and shape their own "sphere publics" through enculturation with folk music rituals, the song circles, performances, recordings, and sonic practices exemplified by these songwriters. For these songwriters, exchange emerges as the central concrete and distinct practice where they trade and re-compose their songs in order modify, record, and perform music for a public audience.

Expanding the Song Circle

After examining the processes operating in the recording studio and in the song circle, how can we better access and study the accomplishments and struggles, the connections and disconnects, and the sounds and silences in each location? We can address issues of social justice and access among songwriters by emphasizing specific themes related to additional identity categories and experiences, including the various "coming out" experiences that queer musicians enact through their songwriting and musical exchange. We can also look more closely at the work of individual musicians who explicitly pursue social justice through their sonic practices and songwriting community engagements. Based on this case study's findings, we can follow up with these experts by conducting personal interviews about additional songwriters' everyday material environments and personal sonic experiences.

Written surveys and scripted interview questions do not mesh well with inquiry into a dynamic event, so additional research could extend the current study, written from Mary's own participatory perspective, to offer more diverse voices, themes, and detailed experiences of individual singer-songwriters. A consolidated approach to follow-up interview questions would broaden an inquiry of songwriting into sonic practices and illuminate diverse cultural ways of working within the musical sonic arts:

1. Are you, or have you been, in a songwriter's group? Describe the experience and what you might gain from exchanging songs with colleagues.
2. When writing a song, do the words, the emotions, or the musical sounds come first? Does it matter?

3. When you write, do you connect to your audience through a shared belief or lesson you're offering in the lyrics? Through an emotion or mood emphasized in the music?
4. Is one of your goals with your own music and teaching to build community? How do you create and sustain community through music?
5. How could you work to diversify your musical community?
6. Do you deliberately try to create and engage the audiences at your musical events? In what ways? For example, do you use strategic silences, call and response moments, sing-alongs, or other techniques to engage audiences?
7. Can you describe a cultural location or event (e.g., a church or a musical concert) where music is central to a specific ethical or political outcome for that culture?
8. How might music best connect people for an ethical or political purpose? When might music offer unexpected obstacles to connection and culture? When can music break down barriers to achieve a specific cultural and political outcome?

We have seen how gathering experiences and analyzing the collective experiences of one small group of acoustic songwriters and producers can offer new snapshots of the sonic practices for various locations. An event like the song circle can be further analyzed by identifying elements of humans' engagements with their environments and also with the listeners around them. Recalling Schafer's analytic terms for soundscapes—keynote sounds, signal sounds and soundmarks (9–10)—we need to recall how these elements are all based in culture. Within music, these elements help mark the characteristics of recognizable musical styles and genres. For instance, when playing backup for another musician, instrumentalists establish the keynote tonality and rhythm as the fundamental background so that the signal sound of an instrumental solo (e.g., a guitar or piano "lead") can stand out. A soundmark, as an identified community sound (like a train's clanging bell), speaks to the culture being preserved. In music, that soundmark can be the continual return to an anthemic vocal chorus or to the song's "hook," for example, in the civil rights anthem "We shall overcome . . . someday, someday." Offering a message that is memorable and meaningful within the cultural context, the music is then performed and sung collectively, so elements can blend and change alongside the audience's sensations of music.

Recognizing and naming the shifts in elements and sensations through resonant listening to music attunes us, as we attend to human communities and connections as well as to the nonhuman environment and agents that surround us. Like designed soundscapes, musical projects offer a window into a composer's choices of musical elements, moods, and styles. However, focusing on embodied responses must also embrace the partiality of that work, knowing that an objective and fully able-bodied approach isn't always possible or even desirable.

Some songwriters have expanded their studio recording projects as interventions into dominant cultural forces. For example, one songwriter recorded a four-song collection in September of 2020, in which she rewrote traditional jazz and pop arrangements after crowd-sourcing the question "why do you vote?" on Facebook. She created new lyrics from those responses that urged people to vote in the 2020 presidential election. Using the affordances of studio recording and her resonance with different musical styles, her release intended to represent diverse backgrounds and offer one small, yet timely, act of social justice. Poised for circulation into culture, her timing just before the election subsequently caused difficulty with distribution into a formal political outlet and, unfortunately, cut off its cultural circulation. Audiences were not able to receive and then amplify her songs. Its moment passed, and it's not clear if this work will be resurrected in 2024 and circulated with an updated message. The means of production, the exigency, and even the outlets for distribution remain uncertain elements in the rhetorical circulation of songs. The evoking of stories and memories through songs, the re-composing of music in performances, and the recordings and their distribution all become the critical variables of that song's life in rhetorical circulation.

Analyzing the singing and playing of music among this group of singer-songwriters shows how exchanging songs remains the most central sonic practice among songwriters. The importance of specific forms of community engagement—with each other and with music—is a new part of understanding oneself as a writer and a songwriter. Communities that form within these shared practices or rituals can generate connections and encourage openness to change. As musical artists, songwriters tend to hold onto the pre-rational, emotional, and embodied nature of their songwriting. Some songwriting colleagues describe their songwriting as the opposite of rational or even rhetorically audience-driven; instead, many describe a whole-body feeling that might become rhetorical later during revisions. Many songwriters describe their process for coming up with songs as happening before any rational thought about it exists. Other songwriters demonstrate how they believe their best songs narrate personal experience in a focused way that can touch on common themes and also still touch other people's emotions. Songwriters use a few common themes for what they believe their song can do for themselves and their audiences: create a feeling/ mood, tell a story/lesson, or offer a gift to others. Songs are often written to teach, to identify with, or to be appreciated by others, but they are also written simply to evoke feelings. Some songs are also written as dedications or presented as gifts. For example, one of the oldest and shyest songwriters in Mary's collaborative group wrote "You Deserve a Song" for a friend and performed it at her birthday party. The song elicited her tears, which he said, "was all I needed to know." In moments like this, songs are exchanged as love, as a central type of human connection.

Playing music with and for other people and doing so in a specific location helps create what Jon Batiste in "Building Community through Music," calls "social music." Batiste shows the audience how community music draws upon the cultural gumbo of New Orleans jazz, folk and blues by playing instruments from each genre. He also demonstrates his "silences" that evoke a call and response moment between Batiste and his dancing, clapping, and singing audience members. Such response, whether explicit or more implicit, is central to the culture of Americana music, where singer-songwriters use sing-alongs, anthemic choruses, volume dynamics, and other techniques in their songwriting techniques. We may not agree with Baptiste's image that music is a universal language waiting in the air, but we can agree that music has both subcultural specificity and important elements of common human experience. Music is emotional, cognitive, and physical all at once—an inspiring example of fully embodied response. It sets moods and sometimes transports us, it makes us feel emotions, and it moves and vibrates our bodies.

Many of the assumptions in this book and our pedagogy rely on the unconscious or precognitive experiences that are possible with sound as a fundamentally physical and vibrational phenomena. Sounds also engender emotions that can evoke perceptions of trust, empathy, and compassion because of their resonance and their cultural significance. Inspired by the sonic practice of singer-songwriters, collaboration, and trust, we can reach toward the freedom described by Batiste and other musicians by helping students to discover their own ways to connect with audiences, even as a kind of momentary transcendence.

Lessons for Student Composers

By way of conclusion, we argue in this chapter that songwriters offer new lessons for all writing—what writing can be like and how collaboration can be successful. We offer these ideas for our students as they create their own sonic texts and other multimodal compositions, both in our classrooms and as lifelong learners.

1. Engaging in music and listening to music both provide an important avenue for telling one's own and others' stories. The first step is listening: learning how to engage in resonant listening with fully embodied response techniques to one another.
2. Immersing oneself in many musical sonic practices and experiences—exclusive to learning a musical instrument—offers healing that can fight against illness, depression, isolation, and pain (Stacy, Brittan, and Kerr). Such immersion can also help students heal and learn to fight against oppression from disparaging people or dominant institutions.
3. Collaborative sonic practices require problem-solving and developing a common vocabulary, then (re-)composing in a way that resonates for colleagues and audience members alike. Students can also use this process to build trust, but it takes time and significant commitments.
4. Song circles can build camaraderie and trust that inspires willingness to risk. Such locations encourage members to negotiate interpersonal dynamics, challenge their own comfort zones, and exchange their original works with other people. Students can follow a similar pattern of exchange and response with their own multimodal compositions.

Borrowing from these kinds of sonic practices, students can gain inspiration, solace, agency, and rhetorical techniques to express themselves and fight back against oppressive dominant cultural arrangements.

The lessons that songwriters can offer students engaged with sonic projects will help them develop careful reduced listening skills and use safe compositional spaces to experiment as they eventually create trusting, long-lasting collaborative experiences. Such findings translate well to other spaces and locations, where engaged sonic experiences occur among sonic art, colleagues, and the environment. The next chapter enacts our sonic pedagogy in an art exhibit and workshop that focuses on echoes among soundscape workshop colleagues, their recordings, artworks, and the land itself.

4 Amplifying the Rhythms of the Land and Environment

[The rhythmanalyst] is capable of listening to a house, a street, a town as one listens to a symphony, an opera.

—Henri Lefebvre, *Rhythmanalysis: Space, Time, and Everyday Life*

While Mary's song circle enacted long-term collaborative sonic practices and relationships, Michelle's case study focuses on a collaborative one-time event—a three-day sound art exhibit—and its role within a larger land archive. The event, named Re/Call, invited artists to create installations that amplified the sounds of the earth with the hope of inspiring participants to challenge the planet's ongoing degradation and, more immediately, co-create an archive of sounds for the Rocky Mountain Land Library, where the event took place. Framing it as a weekend art retreat in rural Colorado, the organizers wished to interrupt what Lefebvre would call the abstract disordered linear tempos of everyday capitalist culture with the more cyclical dynamic rhythms of the earth. By resonating with these cyclical movements and tempos and engaging in the collaborative receiving, amplifying, recording, and exchanging, participants in the exhibit became aware of sound's rhetorical power to create so-called cultural and natural spaces, like cities, rural regions, and conservation and heritage sites, such as the Rocky Mountain Land Library.

After briefly describing the event, this chapter will offer a critical framework, based on Lebvre's rhythmanalysis, for understanding the event's sonic practices as cultural and environmental interventions. The framework includes: 1) critiques of Lefebvre's approach by cultural and feminist researchers, like Lauren Berlant and Emily Reid-Musson, who challenge the linearity of capitalist tempos and the implicit privileges of Lefebvre's rhythmanalyst and 2) the work of Indigenous scholar Kyle Powys Whyte, who advocates cyclical tempos of kinship (vs. crisis reactions) in response to environmental degradation. The chapter will then detail the ways Re/Call drew participants into the diverse rhythms of the ranch, thus building a collaborative dynamic understanding of the region and their kinship to it. Unfortunately, the drive to record and amplify singular "soundmark" sounds for the Land Library's sonic archive was so strong participants failed to record and document the intimate rhythms among bodily movements, the environment, and cultural meaning-making. There was, instead, a rush to isolate and record sounds with high fidelity in order to preserve a postcard memory of the event, thus enacting what Robinson would call "hungry" white colonial settler ways of listening that seek to extract and contain linguistic information for future consumption (2). In this case, participants, many of whom were starved for the rhythms of the ranch experience, sought to both contain and distribute those rhythms. Despite succumbing to this cultural drive toward creating and collecting sounds with postcard fidelity, participants did initially engage in the messy and entangled bodily practices of receiving, amplifying, recording, and exchanging in order to sync rhetorically with the rhythms of the land and its inhabitants. The chapter will detail these highly situated sonic experiences and practices, alongside the more problematic forms of commodification.

Overview of Re/Call Event

Re/Call brought eighteen artists together to create an immersive multisensory art retreat at the Rocky Mountain Land Library, a non-profit residential library located at an historic ranch one hundred miles southeast of Denver. The three-day retreat took place over Labor Day weekend 2018—a time when environmental art exhibits and installations were becoming more common in the metro Denver area but not in more rural regions like the southern part of Park County where the ranch is located. Many of the visiting urbanites camped on the ranch land itself, which the Land Library is slowly converting from an abandoned 1862 sheep and cattle ranch (including a main house, cook's house, horse barn, bunkhouse, corral stalls, and sheep barn) to a residential artist retreat and library supported by a growing collection of natural history books on the South Park region. The high mountain ranch sits just south of Buffalo Peaks and on the banks of the South Platte River, so the weather is fairly volatile, even in late summer, which made the whole outdoor event a bit precarious for everyone involved.

Since most of the installations were sound-based, Michelle was invited to provide a "Composing Soundscapes" workshop, which included a brief lecture on the history and genres of environmental sound art for both the artists and visitors. The workshop was an opportunity to engage in a form of collective audition ostensibly aimed at developing a sonic archive of the exhibit, the ranch, and the surrounding area, but it also allowed participants to experience more intensive forms of resonance and a rhythmic rhetorical reciprocity with the landscape via listening and recording practices.

The ranch's visual archive is already robust (see the Land Library's website and photos by the author below), so the stated purpose of the workshop was to create several "soundmarks" (Schafer 9) to add to the library's visual and print archive. However, these soundmarks were just a small part of a larger environmental workshop objective, which was to encourage visitors and artists to listen deeply to the earth in order to respond to the deleterious effects of pollution, deforestation, erosion, and climate change.

Mary Caulkins, the founder and co-organizer with Louise Martorano of Re/Call, was initially inspired into action by the following statement by Christopher Clark, director of the Macaulay Library's database of animal sounds: "The whole world is singing . . . clicking and grinding and whistling and thumping . . . but we've stopped listening" (*Racing to Extinction*). She hoped participants would "recall" what it was like to listen to the world and "to reflect, celebrate, and memorialize the planet and the jive of its creatures" (1), and Michelle's soundscape workshop was intended to concretize this critical listening practice into the collection and distribution of a digital sound archive. Informed by Lefebvre's critical rhythm methodology, along with Schafer's soundscape analysis, the workshop also invited participants to reflect upon and resist the hegemonic listening practices associated with capitalism's relentless "rhythm of producing (everything: things, men, people, etc.) and destroying (through wars, through progress, through inventions and brutal interventions, through speculation, etc.)" (Lefebvre 65).

Lefebvre's *Rhythmanalysis* and Sonic Rhetoric Research

In *Rhythmanalysis*, his final installment of the *Critique of Everyday Life* series, Lefebvre shifted his attention from spatial analysis to sonic analysis and, more specifically, rhythm analysis with a particular interest in how diverse and conflicting timescales—linear and cyclical, economic and biological—constitute one's experience of the "everyday." Employing a dialectical Marxist approach to the "everyday," Lefebvre argues that capitalism imposes an abstract regimented goal-directed linear rhythm that overlays and seeks to dominate the longer, diurnal cyclical rhythms associated with the earth's rotation, metabolic processes, and seasons. These two rhythms are held in tension, with the struggle itself creating new rhythms, establishing them, then breaking them down again. While more contemporary cultural studies scholars, like Berlant, rightly characterize the rhythms of late capitalism as crisis-oriented and disordered, constantly interrupting and overwhelming the experience of the "everyday" and imposing "rhythms of crisis and response" more than orderly rhythms of daily routine (8), Berlant's critique does not discount Lefebvre's theoretical position—that rhythm and "dressage" (the socializing breaking of an individual into the hegemonic rhythms, whether they be orderly or disorderly) is a defining feature of modern subjectivity.

These hegemonic rhythms are somewhat insidious, infecting even environmental rhetorics with the clock-ticking linear, as well as crisis-response, rhythms of capitalism. In his critique of environmental rhetoric's linear timescales of steadily increasing temperatures and rising sea levels, Indigenous scholar Whyte advocates for the alternative cyclical tempos of kinship. Like Lefebvre, Whyte posits linear time as repetitive, sequential, and unfolding "through the passage of uniform . . . units," while kinship time is based on ongoing cyclical shifts in relationships and forms of rhetorical reciprocity, mutual caretaking, and guardianship (42). The former, depicted through carbon bar charts and sea level models, creates a sense of crisis and urgency and often results in impulsive, scaled-up solutions (mostly market solutions) that keep inequities in place (43). One example Whyte cites is the swift action the Mexican government took to implement wind turbines on the Isthmus of Tehuantepec in the Mexican state of Oaxaca. In order to cut its carbon footprint quickly, the government sidestepped important trust-building negotiations with the Zapotec people over proper land use and profit-sharing. Ultimately, the project failed, due to this impulsive focus on large-scale problem-solving at the expense of the slower rhythms of negotiation, trust, and consent (Whyte 46–7).

Instead of understanding climate change as a ticking clock, Whyte argues for understanding it as an ongoing cycle of creation and disruption in human and nonhuman kinship relations. The colonization and domination of Indigenous peoples over the last 200–300 years, for example, "involved sweeping disruptions in kinship," which then affected "the shared responsibilities tied to the climate system" (Whyte 53). Kinship time, argues Whyte, "focuses us on understanding that kinship relationships are in peril, and we must take urgent action to establish or repair kinship relationships; otherwise, we will not have the interdependence required for responsiveness that prevents harm and violence" (53). Responding rapidly to the ticking clock with wind turbines and other renewable energy systems ultimately fails without the slower movements of mutual responsibility, trust, consent, and interdependency (among diverse humans, animals, plants, and ecosystems), which are necessary to sustain such solutions. With kinship time, solutions emerge from the ongoing (past, present, and future) rhythms of mutual care and responsibility.

To sense these diverse rhythms—the linear abstract rhythms of capitalist commodification and production, as well as the lived kinship rhythms of seasonal change, harvesting, and feasting—Lefebvre emphasizes the central role of the physical body: the body is the "site of contact," a "metronome," or "referential rhythm." for the conflicting linear and cyclical timescales (29–30). His analytical process begins with amplifying the polyrhythms of the body itself: eurythmia (the healthy synchronizing of the beats of the body's organs), as well as an arrhythmia (the unhealthy asynchrony of beats). Though the body of the analyst is central at first, Lefebvre insists that one must be "grasped" by surrounding rhythms in order to grasp them: "one must let oneself go, give oneself over, abandon oneself to its duration" (37). Being grasped by various rhythms requires a particular rhetorical posture or bodily practice of receiving, one that allows the body to receive, amplify, and reverberate with the varying speeds and oscillations of a sonic environment. Most of us human animals discern a relatively limited range of rhythms and often mistake fast rhythms as a constant tone and slow rhythms as a one-time occurrence (Fitch). In his description of the oscillating process of rhythm resonance, Lefebvre claims the rhythmanalyst takes their body's time "as first reference, but . . . the first impression displaces itself and includes the most diverse rhythms, on the condition they remain *to scale*" (46). Many rhythms will remain outside the scale of human perception. "It is therefore necessary," according to Lefebvre, "to situate oneself simultaneously inside and outside"—to remain susceptible to the widest variety of rhythms while also recognizing the limits of one's own bodily sensory processes and perceptions (37). As a social science (and Lefebvre advocated rhythmanalysis as a new science), Lefebvre's approach is almost thoroughly humanized and yet it is an ephemeral, amorphous humanism, with the important acknowledgement that rhythms intersect and complicate easy subject/object distinctions. He stresses that the rhythmanalyst must allow herself to be overcome by the diverse rhythms of her everyday experience. It is an approach that emphasizes an embodied analytical subject (agent) who remembers and records a first impression, but that strong subject must also mark its own inevitable movement toward other diverse rhythm objects: "Observation . . . and meditation follow the lines of force [the rhythms] that come from the past, from the present and from the possible, and which rejoin one another in the observer, simultaneously centre and periphery" (46). With this oscillation, the rhythmanalyst experiences a rhetorical reciprocity—the ability to influence and be influenced—with the land and its inhabitants and understands that even one's own first impression is also within "a rhythm that escapes it . . ." (45).

Lefebvre's detailed account of this resonant oscillation—the repetitive shifts between cyclical, diurnal, and biological rhythms and the abstract tempos of capitalist production in the creation of "everyday" space and time—is what makes *Rhythmanalysis* so useful to social scientists, to rhetoricians, and to this particular case study (27). He offers ways to cultivate: a receiving body aware of its own rhythms and their interactive influence on "outside" rhythms; the ability to amplify certain rhythm currents over others; an analysis that records the repetitions and remixes the interferences of the cyclical, linear, and disordered rhythms that create an experience of a particular place; and the collaboration that allows for the collective audition and exchange of these rhythms toward creating a shared sense of the social and the public.

Much like other critical phenomenologists, Lefebvre provides a detailed method for this type of observation and analysis, which requires one to focus "on separating out the causes and origins of multiple noises, murmurs and clamours . . . by discerning and following rhythms, those of flowers and rain, of childlike or bellicose voices, of secret meetings" (69). Music and dressage enrich his approach with an emphasis on identifying the melodies, harmonies, and socializing rhythms of the everyday into diverse timescales—linear and cyclical. He asks us to amplify, for a moment, one particular sonic repetition, in order to identify its harmonies (synchronous sounds), and, most importantly to Lefebvre, the placement and sequence of the melody's sounds—their tempo. This one particular repetition is then released and organized into the diverse sonic repetitions that stream into and recede from—"flex and reflux"—the scene, which for this particular case study, is the Re/Call Land Library exhibit.

When Lefebvre details his sonic experience of a Parisian day, he is located behind an inconspicuous window—both inside and outside, both "center and periphery" to the sonic currents of traffic, pedestrians, and festivals. Michelle's Re/Call workshop, too, took place on an artisan-crafted porch, inside and outside the diverse rhythms of the exhibit and historic ranch land. The porch was connected to the old horse barn, which blocked but did not eliminate the intense chronic sounds of the northern wind and motor traffic on the neighboring highway. From this perspective, Lefebvre states that rhythmanalysts will soon "see" the garden [the land] "and the *objects* (which are in no way things) polyrhythmically, or if you prefer *symphonically*" (31). He goes on: "In place of a collection of fixed things, you will follow each *being*, each *body*, as having its own time above the whole. Each one therefore having its place, its rhythm, with its recent past, a foreseeable and a distant future" (31). According to Lefebvre, sounds and rhythms offer us a historical perspective, while photographic images enact a never-ending present. "[T]ime is not set aside for the subject," he argues. "An apparently immobile object, the forest, moves in multiple ways: the combined movements of the soil, the earth, the sun. Or the movements of the molecules and atoms that compose it . . . The object resists a thousand aggressions but breaks up in humidity or conditions of vitality, the profusion of miniscule life" (30). Sounds and rhythms are mapped according to time—that is, the expansion and contraction of oscillating waves hitting the eardrum and other parts of the body and, in perception, the arising and falling of tones and frequencies. Because humans perceive and experience them as ephemeral and process-bound, sounds support an historical and social analysis of a place.

Though Lefebvre's analytical object was the French post-war city, geographers and sociologists have extended his methodology to rural and wilderness contexts, as well as the experiences of women, people of color, and immigrants (Reid-Musson). Lefebvre almost unwaveringly addresses only the experiences of white cisgender male urban citizens, so his analysis is severely limited. Cultural and feminist researchers, like Reid-Musson, have rightly critiqued the unacknowledged social and economic privilege that supports his Parisian balcony perspective, as well as the essentialism at work in his conception of the analyst's body and its rhythms. However, Reid-Musson argues, the global rhythms of capitalism do affect us all, though some (particularly impoverished and unhoused people of color) experience them more acutely and violently than others. Everyone, she claims, is subjected to the constrained, often imperceptible (when compared with architecture and the design of the factory floor) daily rhythms of capitalism—almost imperceptible because they are so intricately tied to the "everyday patterning and routines of movement and rest" (Reid-Musson 883).

Reid-Musson puts Lefebvre's rhythmanalysis "in conversation with intersectional theory" in order to analyze the power differentials in her research on migrant farmworkers of color from Mexico, Jamaica, Trinidad, Tobago, and Barbados in southwestern Ontario, Canada (882). She demonstrates how the colonizing state and employers use time and the rhythms (repeated moments of movement and rest) of labor to exploit the migrant farmworkers by subjecting them to long work hours during the agricultural seasons and constraining and surveilling their so-called free time. At the same time, she argues, rhythmanalysis asks the researcher to document the residual rhythms of everyday life, which exceed the constraints of the migration and work regimes (890). She notes the expanded rhythms of Jamaican and Mexican dancing and eating in the evenings during the harvest times and how access to these rhythms was limited to male migrants, while female migrants' "free time" was constrained and surveilled by not only employers but also the migrant community (891). "Gender and sexual exploitation enters into regulating both male and female migrant agricultural workers' access and use of space and time, albeit on different terms," Reid-Musson explains:

> Migrant men's rhythms stand out from the normalized, routine rhythms of white rural Canadian everyday life; the latter includes car ownership, heteronormative families, permanent settlement, participation in paid work from nine to five, and having "roots" in rural communities . . . Migrants' position in rural Canadian contexts is out of place in relation to these normative rhythms of everyday life, in a number of different ways. They move back and forth between Mexico and Canada on a seasonal basis and do not have access to Canadian permanent citizenship; they do not have access to cars, they are racialized, and they live in single-sex dormitories with fellow male migrants adjacent to their workplaces, with limited autonomy over their private space and leisure time." (892)

Migrant women, on the other hand, "face strict feminized constraints from employers and male migrant counterparts. Thus, gender and sexual regulation configures migrant men and migrant women's relationships to one another; migrant women are not only monitored by men but can be dependent on them outside of work in order to circulate and gain access to services and social spaces like church or simply going out for coffee" (892). Reid-Musson also explores the ways her white Canadian female body limits her access to spaces and time within the migrant work and leisure spaces. In challenging the essentialism inherent in Lefebvre's researcher-body, Reid-Musson demonstrates how the a priori categories of gender, race, and class are constituted and challenged within the rhythms and analysis of everyday life ("the processual and repetitive patterns and routines") (893).

Re/Call's Receiving and Amplifying Practices

As Lefebvre suggests, experiences of the ranch, the porch, and Michelle's workshop itself were preceded by a series of sonic rhetorical engagements and rhythms, and as Reid-Musson claims, these engagements and rhythms contained both challenges to and support for gender, racial, class differences. Because rhythm is repetition and pattern, and therefore inherently historical, this section will situate the resonant listening practices of the workshop within the larger collaborative planning and organizing rhythms leading up to the event. When she began organizing Re/Call and writing the call for proposals, Caulkins met with Michelle to talk about research on sonic rhetoric and climate change and how she might persuade people, particularly white middle- and upper-class urbanites, to stop, listen to, and intervene in the disordered and disorienting rhythms of consumption and environmental degradation and disaster. Caulkins also wanted to create a feminist maker space for exhibits by people marginalized by the dominant art culture, such as women of color and LGBTQ+, as well as people who did not see themselves as artists but loved the land and wanted to memorialize and value it together in some way. The art retreat was meant to be an immersive experience built into the cultural (albeit imaginary) "pause" in productivity associated with Labor Day weekend, and the call for proposals made this clear:

> This retreat is seeking to create an experience that both celebrates and memorializes our earth, its sounds, its beauty and its sacred resources both tangible and ethereal. Artists are welcome to propose performances, participatory experiences, projects, lectures, installations, sound art, film or any array of work that engages and perhaps amplifies the multi-sensory experience of our natural environment. (Caulkins and Martorano)

As Caulkins and Michelle continued to discuss the goals of the retreat, the affordances of sound for intimately experiencing the environment became the central focus. Caulkins was particularly interested in the argument Mary and Michelle developed on environmental sound art and the nostalgia and "collector's exuberance" behind the recording and cataloging of sounds from extinct or endangered species.

Ultimately, Caulkins decided to incorporate this drive to collect, alongside a desire to reverse destructive actions, into the exhibit's Invitation for Proposals: "The title of the project, 'Re/Call,' was formulated through the reflection of goals for this three-day experience. The first being the nostalgic recollection of the resources that are disappearing and the desire to 'recall' or reverse the actions that lead to their depletion." Caulkins wanted participants simply to slow down during the long weekend and enjoy intimacy with the land: "Re/Call is about taking time to enjoy this planet because many of us spend so much time hustling and only rarely stop to reflect and recall the resource that we all share. This project seeks to explore our creative options to both enjoy this world and disrupt that trend. It is a time to enjoy each other in nature over some time." From the start, Caulkins set the rhythm of the project: a slow beat, which allowed time and space for the artists to gather and co-create the retreat and exhibit space. Once Caulkins and her co-organizer Martorano selected the key proposals, including Michelle's, for the event, the artists all met regularly on the ranch and over meals to attune their individual projects to the land, the growing land library project, and each other. During these early visits, Michelle focused on cultivating a body and tools (microphones and playback devices) capable of receiving and amplifying a diversity of ranch sounds and rhythms.

Michelle's first "receiving" visit to the land with the other artists occurred nearly a year before the actual retreat. From the beginning, Caulkins and Martorano wanted artists and presenters to spend large swaths of time with each other on the ranch learning about the land and the library itself in order to take a break from what anthropologist Anna Tsing calls the "driving beat" of modernization. "Without that driving beat," Tsing writes, "we might notice other temporal patterns. Each living thing remakes the world through seasonal pulses of growth, lifetime reproductive patterns, and geographies of expansion. Within a given species, too, there are multiple time-making projects, as organisms enlist each other and coordinate in making landscapes" (21). Throughout the sixteen-month Re/Call project, the artists learned about the history of the landscape, the ranch, and the library from co-founders Jeff Lee and Ann Marie Martin. As they slowly moved across and through the land, Jeff and Ann would detail the animals and structures they initially found there, pointing to discarded metal roofs and ranch equipment, ripe for salvaging. They shared how many of the ranch's original buildings were burned in a fire created by a passing coal train and how Re/Call, among other events, was part of the larger recuperation and survival processes following this early fossil fuel disaster. Though the rhythms of the storytelling and collective and individual meanderings around the ranch were slow and cyclical (based on the movements of the earth and the daily monsoons) relative to daily work lives, the disordered rhythms of production and distribution still punctuated the "everyday" with large tanker trucks barreling past the ranch at random intervals transporting fracking oil from farms to cities and with workers demolishing walls and removing debris as they rebuilt the cook house, a large six-room wooden cabin still damaged by fire, water, and many generations of deer mice. The retreat would make use of the actual ranch land for participants, clearing out debris for tents, camper trailers, and a portable kitchen.

The main rule during all the visits to the library was no paper waste. Everyone brought in their own cups and plates and carried out any trash (wrappers, water bottles) they created. Participants would, therefore, leave the ranch with the weight of their own consumptive waste. Of course, the waste was eventually funneled to another part of the land, but the experience of carrying one's own trash—however brief—made visible the regular consumptive and destructive labor (by the land and others) of waste management, and participants were made aware of the impact of their daily, often hourly, repetitive waste practices on the living environment. While modern community is often organized around comfort—a ritualized ballast against natural elements—artist visits to the land and the retreat itself took away some of those buffers and exposed those who chose to camp or spend even an afternoon on the land to intense rapidly shifting currents of cold and wind compounded by a jet stream made more erratic by global warming. The discomfort and sickness that arose from that exposure laid bare the amount of labor and resources required for human life on the ranch and the effects of that labor and extraction on the climate itself.

Michelle spent her early visits to the ranch engaged in receiving and amplifying practices and, following Lefebvre, created a body receptive to the sounds of the ranch and bent toward a reciprocity with the land as she walked and rested. She noticed, amplified, and recorded particular sound waves that arose and decayed—what Schafer calls sound "envelopes" (129)—and their repetition with other sounds in what constituted a rhythm. She tried to discern and isolate the birth, growth, peak, and decay of particular bird sounds; the wind moving through discarded metal ranch equipment; barn doors and cattle gates slamming; the river rushing over rocks and through grasses; the more subtle sounds of sagebrush moving across the dirt; the crunch of dried grass beneath her footsteps, as she dodged fox and rabbit holes; and the disordered sounds of her own labored high-altitude breathing. Those were the key rhythms she chose to amplify with her initial recordings. They were the rhythms that told her she was at the ranch. Over lunches and tours of the ranch, she learned many of the artists—most of them women—were creating forms of sound art for the first time and did not really know how to situate their work within a growing tradition of sound art or in relation to other work on climate change and environmental degradation. She then understood her own workshop as an opportunity to amplify their work in relation to this growing tradition and as a space for them to reflect on the impact (on their own bodies and the land) of their work.

In terms of the "final" installations for the exhibit, the many individual and group visits to the ranch and the sixteen-month collaborative and creative process culminated in three soundwalks: one along the river called *Chime Walk* (Hall), one along a bird flight route called *Four Songlines* (Scott), and one on the eighty-mile drive from Denver to Buffalo Peaks Ranch called *From Here to Where* (Burgund). The collaborative process also produced one immersive virtual reality dreamscape set up in one of the barns via hammocks and high-tech headphones (Wolf); one sound sculpture (Barbee and Nelson); and a replica of an ancient Indigenous amphitheater, where retreatants could go sit, listen to the landscape, and inscribe their experiences or thoughts on earth bricks, which were then added to the replica (Tepp). All were a form and instance of rhetorical reciprocity with the land and participants, where they collectively experienced the rhythms of the ranch via birdsong recordings and musical and architectural instruments. These rhythms were then recorded and became part of the history of the ranch or, rather, the history of place-making on the ranch.

The Soundscape Workshop

Michelle's initial proposal for the exhibit/retreat was a lecture focused on a critical discussion of The Cornell Lab of Ornithology: Macaulay Library of Sounds, which was also Caulkins' inspiration for the retreat. She wanted to showcase how artists have made use of the library for environmental change and at the same time discuss its limitations as a largely decontextualized database of isolated sound signals. When she first visited the land with the other artists, however, she became more interested in conducting a workshop on composing soundscapes. The rhythms of the ranch and its inhabitants imposed themselves, asking to be received, amplified, recorded, and exchanged, so she started developing a participatory workshop, which would prompt participants to tune into the rhythms of their own bodies, using Lefebvre's metaphor of the analyst's body as metronome, and to the rhythms of the ranch landscape (e.g., the river, sagebrush, hills, the creaky cattle gates, barn doors, and the relentless rapidly shifting high country wind). The culmination of the two workshops would be a sonic archive of the event and the larger rhythms and sounds of the ranch posted on the Re/Call and Land Library's web sites for public access. Specific information and directions from the handout she developed and distributed to workshop participants are below.

Composing Soundscapes Workshop Handout

"Everywhere there is interaction between a place, a time, and an expenditure of energy, there is rhythm."

—Henri Lefebvre

A. Brief Overview of Environmental Art/Sound Ar

Category	Example
Sonic Landmarks	John Morton's "Central Park Sound Tunnel"
Sonic Maps	Maya Lin's "What is Missing?"
Sonified Climate & Environmental Data	Brian House's "Animus"
Soundwalks	Janet Cardiff's "Wanås Walk"
Amplified Animal "Voices"	Michael Thomas Hill's "Forgotten Songs"

B. Listening Practices

1. Reduced Listening: We listen for the birth, growth, peak, then decline and end of a sound object, as well as its repetition or rhythm. Borrowed from sound artist Chion, this practice asks us to "bracket" our impulse to identify the sources and meanings of the sound in order to understand its physical qualities. Since completely bracketing our meaning-making impulses is difficult, if not impossible, we can observe how those processes guide our perceptions and experiences of sound.
2. Embodied Listening: Next, we lighten our focus on the sound object and notice how we are reacting to and making sense of the objects in our bodies and minds. Sound artist Pauline Oliveros asks us to notice where we feel the sound in our bodies and what our physical sensations are when we hear the sound. We then note the "feeling quality" of our awareness of each sound and the larger soundscape.
3. Ecological Listening: We then engage in an ecological (rhetorical) listening practice focused on how the inhabitants of a partic-

ular place influence one another via sound. Arguing that "the general acoustic environment of society can be read as an indicator of social conditions which produce it," soundscape pioneer Schafer asks us to listen for the "keynote" sound (a background or fundamental tone against which other sounds are perceived, such as the sea) of a place, as well as its "signal" sound (the foregrounded sound to which the attention is particularly directed, such as a boat whistle) and "soundmark" (a community sound which is unique or possesses qualities which make it noticed by people in the community, such as church bells).

4. Cultural Listening: Using your smartphone, record what you consider a keynote sound, signal sound, and soundmark on the land here at Buffalo Peaks Ranch. Don't try to mask the sounds of your own bodily movements or vocalizations since you also want to record your own sonic interactions with the environment. This is an interpretive exercise, so not everyone will record the same soundmark (though there does tend to be a high level of social agreement about what sounds are important). Group reflection on recordings: *What does your recording tell us about environmental, social, and cultural rhythms and conditions of Buffalo Peaks Ranch? The South Park region*? Please email me your sound files, and I will compile them into a sonic archive of the retreat. Thank you for your participation!

Those early recordings and conversations with the artists prompted Michelle to create a participatory workshop that would train people to record the rhythms (and melodies and harmonies) of the event but, more importantly, train them to amplify their own concrete human interactions with it, including the ways their behaviors and movements impacted the local environment. Her first workshop was in the early afternoon, after the opening ceremonies and before the artist tours (and the impending late afternoon storm). It was the perfect time to become acquainted with the land and the art.

A group of fifteen people, including several of the artists, arrived at the horse barn porch (see Figure 7) with phone recorders in hand. Like Lefebvre's Parisian balcony, the porch allowed access to both "inside" (with a sound wall that made tuning into one's body's rhythms possible) and "outside" (with direct access to most of the installations, the river, and large open field and its surrounding ridge). From the porch, they could distinguish sounds. They were not immersed in them, nor were they listening strictly for the purpose of orienting themselves and navigating a field. One does not need a "transparent window" to access the rhythms—it comes through the walls. In fact, though Lefebvre calls the point of observation an "invisible window" (36), the metaphor is inadequate for its focus on the visual. It does not account for the complexities and the diversities of sensing, experiencing, and making meaning of sound. From the array of their own body's rhythms, which is more complex than any window, the rhythmanalyst "will listen to the world, and above all to what are disdainfully called noises, which are said without meaning, and to murmurs [rumeurs], full of meaning – and finally he [*sic*] will listen to silences . . . He listens—and first to his body; he learns rhythm from it, in order consequently to appreciate external rhythms. His body serves him as a metronome" (Lefebvre 19). After the workshop participants settled on the porch, then, the first move was to cultivate a sense of these diverse bodily rhythms.

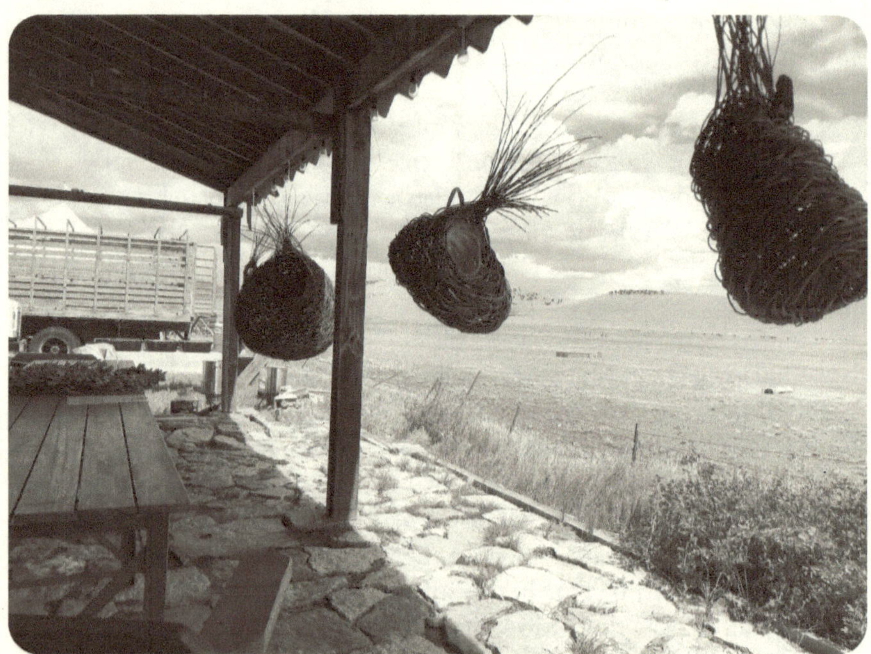

Figure 6: Horse Barn Porch. Photo by the author.

Sometimes the body's rhythms are more discernible through arrhythmias, when the body's rhythms are different from those outside it, even temporarily. Thus, Michelle provided workshop participants with a "guiding sound"—the bell of a gong—in order to help them develop a sonic attention to the rising and falling of the bell's envelope and to its rhythm against those diverse rhythms of their own breathing, pulse, and digestion (the more prominent rhythms in some, not all, bodies). Then she played the famous "Etude aux Chemins de Fer, 1948" recording by Schaeffer and asked them to attend to the recording's rhythms as they arose and interacted with those in their bodies. She chose Schaeffer's recording of train sounds (steam engine, whistles, wheels on tracks) partly because it alluded to the coal train that used to stop right at the ranch's gates in the latter part of the nineteenth century. It was an early example of the linear rhythms of industrial capitalism intersecting with the cyclical agricultural rhythms of seasonal and diurnal practices.

After asking participants to receive and describe the physical sounds in the train recording without resorting to making meaning or dwelling on the identification of a source (a form of reduced listening described in more depth in chapter two), Michelle invited them to reflect on the impact of the sounds on their bodies in order to construct the "metronome" body Lefebvre describes as the center of his analysis. Where in their bodies did they feel the sounds? What sorts of emotional responses or intensities arose and where did they feel those rhythms (whether eurhythmic or arrhythmic)? Several found the sounds hypnotic, with the low frequency linear rhythms of the early nineteenth century train engines punctuated by porter whistles. Again, she avoided discussing any larger cultural, racial, or gendered meanings or patterns associated with these sounds and instead focused on the physical qualities of the sounds and their physical and emotional impacts on bodies.

Once participants developed this form of embodied attention, Michelle trained them for receiving and amplifying the rhythms of their bodies alongside those on the ranch. They practiced recording sounds with their phones in the outdoor classroom and reflected on their physical qualities, then she gave them three sound categories from Schafer's soundscape framework: keynote sound, signal sound, and soundmark (9–10). This was simply a focusing and organizing tool for discerning as many rhythms as possible in the surrounding area. She wanted them to record a large variety of sounds as they walked through the landscape and art installations, but then she wanted them to listen together to their own recordings. Finally, Michelle wanted them to amplify, reflect on, and remix them collectively through Lefebvre and Schafer's frameworks. The soundscape heuristic is admittedly reductive—reducing an environment to a taxonomy of rhythms—but the collective recording and listening does allow for the exchange of multiple diverse soundscapes. Though the orienting drive is often to identify the overriding soundmark of a place, most often there isn't just one.

Because the ranch landscape is so visually stark and vast, Michelle was surprised by the variety of sound recordings participants brought back to the porch. Many were rhythmic with the tempos of wind and rain cycles. One participant recorded the wind blowing through discarded steel roofing in an otherwise "silent" remote field of the ranch. When he amplified the recording back at the workshop, participants reflected on the physical qualities of the sound—the high frequency friction between the air and the metal and the rhythmic banging of one piece of steel on another. They discussed how they experienced these rhythms in their bodies. One participant found the sound of the clanking metal reassuring against the silences created by the wind. Another participant found it produced melancholy but only after she projected the larger meanings of waste and environmental degradation onto it. Despite her sympathy for this interpretation, Michelle steered the group toward the first response and emphasized the reassuring effect of the "impersonal" rhythms of human-made objects interacting with nature. She wanted the group to understand and possibly feel collectively the drive toward industrialization, toward organization and development, toward the creation of distinct soundmarks against the ambient omnipresent noise of the wind, toward beauty and company on the visually stark landscape. Another participant recorded her rhythmic footsteps and strokes of the chimes of the soundwalk installation by Nathan Hall along the river. Of course, more people experienced pleasure with this recording, with the footsteps and chimes connoting artful objects and backyard territories in U.S. culture. The effect was felt through the cultural and class rhythms of domestication and safety along a river rushing with afternoon rains.

Once participants reflected on their embodied experience of the recordings, they began categorizing them within Schafer's cultural framework. They decided the banging steel in the field was a signal sound, strongly directing people's attention toward it when it was present. It was not a random sound but responsive to the cyclical wind currents, which would increase in speed and intensity as the day went on. The chimes, too, were a signal sound, reminding people they were part of an art installation/retreat and the larger domestication rhythms and projects on the ranch. Over the course of both workshops, they ultimately identified the creaking iron gates, in need of repair and banging in the wind, as one of the ranch's soundmarks. Most of them located themselves immediately at the Rocky Mountain Land Library when they closed their eyes and heard the gates. Keynote sounds were the wind (almost ever present and a character in its own right), along with the bursts of recreational and oil truck motors along the neighboring highway. This collective audition and exchange process helped them identify the conflicting rhythms of the two keynote sounds. Using Lefebvre's critical framework, they noted the intermittent disordered time scales of the oil trucks.

Though more common at sunrise and sundown, the oil trucks would blast by the ranch, barreling down the road over the speed limit in order to meet the daily quota. Only when behind a sound wall would one experience these disordered rhythms as background noise, as normalized within the larger ordinary noises of auto-dependency. More cyclical were the rhythms of the recreational traffic, representing a break from the strictly routinized timescales of the workday. Interfering with these motorized keynote sounds was the rushing rhythm of the wind. Because it was often relentless and seemed free of pause or rhythm, the wind sometimes wreaked havoc on bodily rhythms, which were more adapted to windless urban landscapes, and drowned out other more subtle human interactions. Its rhythm, when discernible, was cyclical, with large intervals from dawn to dusk and moving from keynote to signal and back again. Both the capitalist timescales of the oil rigs and the cyclical timescales of the wind framed the experiences of the installations and interactions with the land.

Nearly all the participants also recorded and identified the sound installation *The Sound Mirror Project* as one of the soundmarks for the Re/Call event. Created by artists Libby Barbee and Bill Nelson, a sculpture of assembled organ horns (resembling a sound mirror) amplified and reflected the sounds of calves crying after being separated too early from their mothers on a nearby ranch (see Figure 8). The social reference to immigrant children being separated was readily apparent to most participants in September of 2018. Because the apparatus was connected to a solar power source, the cries were cyclical with the appearance and disappearance of cloud cover and nightfall. At the beginning of the workshop, participants tried to bracket the social and environmental context for the piece. Barbee and Nelson had conducted extensive observations of and interviews with rural Colorado farmers and ranchers, who identified meaningful sounds from their land and expressed ambivalence for the big food processing and high-speed production systems that forced them to separate the calves too early in their lives (5). The wearable and displayed apparatuses they constructed to emit and amplify these sounds are inspired by the images of the acoustic mirrors used by the military to capture, reflect, and focus the sounds of both friendly and hostile ships and aircraft (4). Barbee and Nelson hoped the sound mirrors would help city dwellers locate the sounds of small rural farmers and ranchers in the fog of an urban-focused media and politics in the state of Colorado (5). While the mirrors did not technically reflect the sounds of the ranch, they did reflect and distill a larger rural soundscape associated with the practices of the food industry. For the workshop participants, simply identifying the rhythms of the wailing calves as they walked the land, then recording and amplifying them for each other created a visceral effect on their bodies, a recognition of grief for and with the animals making the sounds—animals that were violently displaced and yet an integral part of the ranch and its past rhythms.

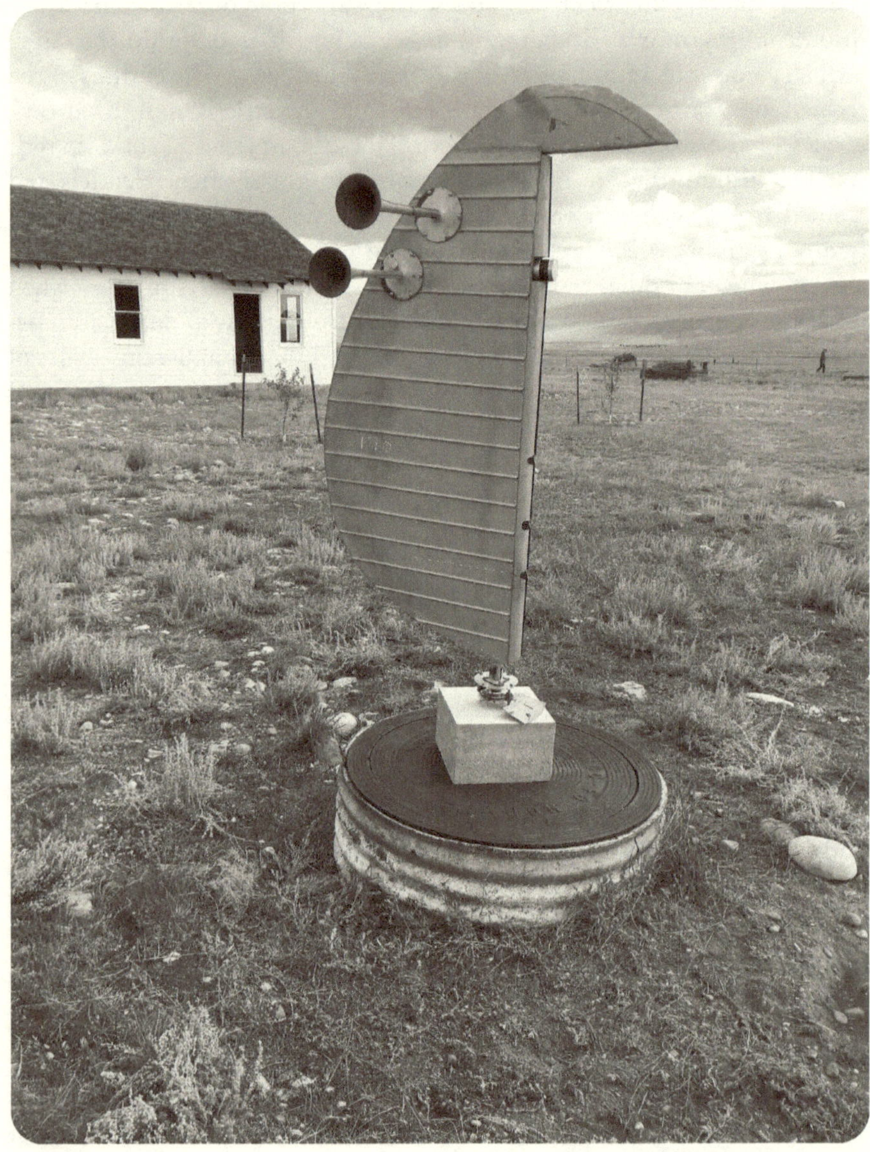

Figure 7: *The Sound Mirror Project* by Libby Barbee and Bill Nelson. Photo by the author.

The Drive Toward Fidelity as Cultural Dressage

After the participants exchanged their recordings with each other, they stored them on their phones and emailed them to Michelle after the workshop. By the time they emailed, many had already remixed the sounds into a sequence, selecting particular bits and stringing them together. Most added recordings of the clanking gate, the sound mirrors, the chimes along the river, and the crunch of their footsteps in the sagebrush, as their desire to capture and archive their experiences of the retreat sonically and with some degree of fidelity grew over time. Michelle, too, wanted to hold on in some way to the intimacy the art and workshops created with the land and the ranch and the way the sounds and their exchange located her there in community with others. In his rhetorical study of the Lomax archive and African American folksong in the 1930s, Stone argues for sonic archives as a space for theorizing new methods of historiography (6). He claims sound has "a powerful historiographical multi-tonality that communicates complex historical experience" (9) and advocates methods that allow one to hear and write "a dissonance that more accurately reflects the various contradictions of our everyday experience" (9). Sonic archives especially confound our compulsion toward historical narrative accuracy and fidelity. As retreat participants compiled and remixed their recordings into sequenced archives then emailed them to Michelle, the sounds became more and more concrete and less contextualized in the collective human experience of the land and exhibit. The drive to capture singular soundmarks and signature rhythms exemplified Lefebvre's notion of dressage—a normalized hegemonic way of listening that overrides the dynamic and cyclical rhythms of the body, the community, and the environment. In this case, participants failed to recognize and record human interactions, the everyday sonic experiences on the ranch, the rhythmic rhetorical reciprocity, and the ways of listening that made these sounds possible and meaningful.

For example, one participant emailed a recording of the "Native Opening" by Regina Lopez-Whiteskunk and Selwyn Whiteskunk, both members of the Ute Mountain Ute Tribe (the Weminuche Band of the Ute Nation). Their tribe's traditional territory spans the western side of the Continental Divide in Colorado, Utah, Arizona, and New Mexico and includes the Buffalo Peaks Ranch area. The organizers had invited Lopez-Whiteskunk and Whiteskunk to open the art exhibit retreat with a traditional prayer-song, acknowledging and praising the Weminuche ancestors who once lived together on the land creating and protecting vibrant cultural and kinship ties with each other, members of other Indigenous communities, as well as the larger ecosystems that sustained them. Only a small group of artists and participants at the retreat identified as Indigenous, so the vocals and rhythms of the opening prayer-song in the Ute language were remarkable to many and stood in "counter-rhythm," as Reid-Musson would suggest (893), with habitual community and individual daily rhythms.

In some ways the prayer-song served as a sonic land acknowledgement, rhetorically intended to remind participants of the genocide white settlers engaged in against the Indigenous nations and communities. The prayer-song, however, also served the ceremonial purpose of officially opening the installation/retreat, a rich form of invoking rhetorical reciprocity and kinship among humans and animals that had inhabited the land before and now. Whether the mostly white settler city-dwellers heard the prayer-song as an invocation and felt those kinship connections during the opening ceremony or at any point in the retreat is unknown. The participants in the workshop who recorded the prayer heard it as a sonic historical marker, as part of the ranch's sonic archive collection, as a nearly extinct sound of the past, rather than an invocation of kinship among past, present, and future inhabitants. Whether or not the participants felt or understood the kinship relations invoked by the song, most participants felt through the body and its senses the paired phrasings, five-part rhythms, and breathing techniques of the Ute prayer-song, which created a strong opening rhythm for the art retreat and a counter-rhythm to the loud monotonous hums of commuting cars and generators. It is possible that some were even aware of the ways their own breathing rhythms synched with and/or resisted those of the singer's, bringing them into literal autonomic relationship with each other. Their bodily responses were, of course, at least partially informed by cultural conditioning—that is, whether their bodies heard the song as noise, melody, story, or timbre and whether they felt themselves participants or spectators in the ritual.

The Ute prayer-song did not carry a driving percussive rhythm; instead, the rhythms were shaped by the phrasing, the shifting pitch, and breathing techniques of the singer. Thus, the synching of the listener's body's rhythms with the song occurred subtly and slowly at the level of breath, as the melody moved repeatedly from the higher to lower registers and from louder to quieter frequencies. The song challenged the linear, compulsive goal-driven directional rhythms of the commuting cars and trucks and even the rhythms of people arriving at the retreat, setting up the tents, and guiding the visitors through the exhibits and through the food lines. The cycling rhythms of the melody as it moved higher and higher in pitch before descending then beginning again, without a clear sense of ending or envelope, brought participants into a more circular, less directional, rhythmic reciprocity with others and their surroundings and introduced what Lefebvre would call a "polyrhythmic" interruption to the normalized rhythms (the dressage) of the everyday (16). The song created a cyclical slowed-down vibrational pattern for interactions at the opening ceremony and beyond as it resonated with movements toward the workshop and other work and leisure activities throughout the day. Although many participants, including Michelle, were "on the clock" that weekend—that is, engaged in paid or donated labor and following a strict workshop and exhibit schedule—the rhythms of the Ute prayer-song interrupted any regimented boundaries between work and leisure and brought the rhythms of kinship (e.g., eating, making art, inhabiting the land together) to the fore. The recording conveyed the distinct nuance of Lopez-Whiteskunk and Whiteskunk's individual voices, given resonance by their bodies and histories, which is what makes a sonic archive compelling. Stone highlights the recorded "voice's uncanny ability to connect us with the human" (91), but the "presence" of that voice is interrupted by the reproduction, the recording apparatus, which "reminds us of the disconnection and temporal distance" between the human bodies and the recording (91). As Stone puts it, the recording renders the human a ghost.

In Kim Shively's video of the Re/Call event posted on the Land Library web site, Lopez-Whiteskunk emphasizes the unrecorded and unacknowledged ghosts in the landscape and offers more context for the opening song—a context missing from Michelle's archived collection of found sounds. "When was the last time the sound of a Ute voice was heard in this valley?" Lopez-Whiteskunk asks. "Land use. What does that mean? What did it look like then? What does it look like today? And how do we get along with one another in those respects? And those conversations—how are they intersecting?" The gesture toward missing voices and kinship ties appeared again later that same day when artist Gregg Deal, a member of the Pyramid Lake Paiute Tribe, performed a Pow Wow dance entitled "Invisible Loss" with his daughter Sage with only the sounds of the bells and trinkets on their black clothing to mark the movements of their bodies against the landscape of the ranch. Wearing black instead of the usual bright traditional colors of Pow Wow dancing and wearing headphones instead of amplifying the traditional music was, as Deal described it, "a metaphor for existing, but not existing" (12). The dancers heard the music through the headphones, but all the audience heard were the sounds of the black outfits. "It's like, you can witness this loss," explains Deal, "but you don't get all the privileges that go along with it. We are omitting the sounds of Pow Wow music and all the color, and when you eliminate those things, then it creates something different than what an American may expect to see. The idea is to say, we are here, but we're not here" (12). Occurring toward the end of the opening day, Deal's performance with his daughter was in counter-rhythm with the opening Ute song. It was nearly impossible to amplify, record, and, thus, archive the muted sounds of the trinkets as Deal and Sage danced to the sounds of their own music. Participants heard only traces, the aftereffects, of the dancers' bodies moving in sync with the rhythms of the music, and the sounds themselves became indiscernible—detached from source and meaning—on a recording. The resulting literal and abstract silences were resonant with loss. Against the hegemonic sounds of traffic and the daily sounds of redevelopment at the ranch, the trace rhythms of Indigenous cultures and nations were barely noticeable and indecipherable without the visual regalia, instruments, and voices. That was the point. The performance cut through attempts to capture and archive experiences of the land through soundmarks, keynotes, and signal sounds and rhythms by making the silences and counter-rhythms much more salient.

Another important audio performance stood in contrast and counter-rhythm to the small sonic archive—artist Halsey Burgland's "From Here to Where," an audio app that augmented the commute from the city to the ranch. Though more ephemeral than the archive's database of sounds, the app documented and integrated the sounds of humans, animals, natural elements, communities, and cultures along the major highways. Because Burgund's piece adjusts to the rhythms and pace of travel, listeners heard a different soundscape depending on whether they drove fast or slowly or whether they stopped along the way. The affective rhythms of the piece changed as well, depending on the pace, the curves in the road, and the altitude, which drew attention to how various rhythms were rhetorically influencing one another. Those rhythms included the driving linear rhythms of the car, the more cyclical rhythms and textures of Burgund's ambient music and field recordings, and the rhythms of one's own body breathing and listening from the car seat or the side of the road. Burgund wanted his soundscape, which was textured with extinct sounds of wildlife and birds, to contrast starkly with the driver's soundscape. Before the show, he said he hoped "to make a comparison between that physical journey and a trip back in time to an earlier moment, when human beings didn't have quite the same influence over the planet" (8).

Although Burgund's app, with its ephemeral reliance on concrete movements through space, would not normally be considered an archive, it does, in fact, create a powerful rhetorical experience of the past. While traditional archives also ask us to engage physically with texts and media in order to create a past, Bergland's archive makes the body's own rhythms and movements an even more prominent part of and a stronger rhetorical influence on a past that is created through interaction. The sounds and beats of birds and wildlife interrupt the homogenous driving linear rhythms of the commuter car that is going somewhere, as well as the traditional road map and GPS cartographies that emphasize roads and routes over the habitats and migrations of animals. Instead, Burgund provides the complex rhythms—some extinct—of multiple species influenced by the speed of the listener's vehicle within what sound artist Samuel Thulin calls a "cartophony" (192). Burgund's app does more than simply amplify the loss of extinct species from the landscape; it demonstrates the loss of sonic diversity in terms of rhythm, pitch, timbre, and frequency in everyday experience. It is a sonic diversity that relies on kinship ties and interactions among humans and nonhumans.

What was missing in Bergund's app and in the event's sonic archive, though, was the sound of the archivist/artist or, more specifically, a vocal, rhythmic, or vibrational trace of the archivist's influence. Recording such influence would have offered a richer sense of the kinship interactions with the landscape and its inhabitants, which made the archiving possible. Sounds of the human body and its rhythms signal a whole body interacting with its environment. Of course, this signal is cultural. If one has learned to associate a voice on the phone or on a voice recording with an actual person, then when one hears a human voice or the sounds of a body in the environment, one is more likely to imagine a whole human (vs. a disembodied voice) walking through dry grass, moving debris to redirect the wind, and moving in conjunction with others through a particular exhibit. Without a visual representation, one may be even more likely to imagine a whole human. Again, cultural differences will inform whether one imagines a cisgender woman, a transgender woman, a cisgender gay man, a Black transgender man, and so on, but freed from a visual representation, one can imagine an acoustic space inhabited by a whole body (or bodies) interacting with or deeply embedded with the environment. Recording and isolating the soundmarks does little to create an interactive acoustic space, a larger space that makes room for kinship interactions among bodies, plants, and the elements. The focus on the fidelity of the barn gate or the bird's song literally extracts it from the complex environment of the landscape, erases the human rhythms and movements required to record it, and reinforces the notion that humans are largely separate from nature and the natural environment. While the high-fidelity sound of a bird's song can be useful in amplifying voices that are endangered and often absent from decision making about land use, such focus serves to not only fetishize the song itself but also to homogenize the landscape.

Ecological Recording Practices

In retrospect, a second workshop that asked participants to capture their actual engagement with the rhythms of the land, whether through spoken words, nonverbal sounds, movements, dance, and/or music would have provided them and the Land Library with an archive of not just sounds from the land but an archive of listening and responding to the land, an archive of the interacting and interrupting rhythms of kinship. The soundwalks of Westerkamp and Cardiff would have provided examples of an artist's sonic and embodied engagement with a landscape. Westerkamp's "Kits Beach Soundwalk" uses the artist's voice to record her push-pull interactions with the overwhelming homogenizing sounds of the city as they drown out the tiny sounds of barnacles feeding, her own footsteps on the beach's surface, as well as the sounds of her own thoughts and dreams. Sound arts scholar Cathy Lane praises Westerkamp's feminist challenge to the "voiceless" soundscapes that were more common at the time, claiming Westerkamp "uses her commentary to suggest an unbalanced world where individual voices are silenced by an imposed authoritarianism manifested through sound" (7). Lane also challenges sound studies scholar Brandon LaBelle's argument that Westerkamp creates an unnecessary and simplified binary between city noise and natural sounds. She claims, instead, that Westerkamp is actually "making space for the small sounds of domestic activity" and by doing so, "a wider range of being and belonging in the world can come into being" (8). Rather than simply rejecting the noise of the city, Westerkamp states, "The city is too loud to listen to the small sounds . . . it interferes with my listening. It occupies all acoustic space . . ." Throughout the piece, Westerkamp amplifies the sounds of the barnacles, the wind, her footsteps, and her internal dreamscapes within and alongside the sounds of the city. She, thus, performs the role and perspective of Lefebvre's rhythmanalyst, who is both "centrecenter and periphery" to the sonic currents of the environment and attuned to the synchronous and asynchronous rhythms of the body and its surrounds.

Despite their failure to fully capture or record this important perspective, participants in the workshop did ultimately engage in what Lefebvre calls "the push-pull exchange of the general and particular" (ix), receiving and amplifying certain sounds over others as they tried to make sense of and cognitively map the new terrain and its complex subtle sonic forms of environmental art. As soon as they all arrived at the retreat, participants engaged rhythmically with the land and its high-altitude movement toward day and night and the seasonal rhythms of afternoon storms and evening cold. The exchange continued with the recording, saving, and playing of sounds with other participants, as they collectively composed an archive of soundmarks of the ranch and the exhibits from their disparate recordings. With each playback, they felt they were knowing the land and building a keener understanding and orientation toward it. Congruent with Lefebvre's argument, the recordings and their exchange changed participants' perspectives on their surroundings. They understood that there was "[n]othing inert" on the ranch—"no things; [only] very diverse rhythms, slow or lively (in relation to *us*)" (17).

Claire Revol, a leading French Lefebvre philosopher, argues that rhythmanalysis constitutes an "urban poetics" of lived space-time that refuses abstraction. The soundscape work in South Park likewise reshaped the rural and cultural space of the Land Library. For three days participants recorded and remixed the repetitive uses of space, time, and energy—what Lefebvre called "rhythm." They recorded the cultural rhythms of tourism, of exhibition, of fossil fuel extraction, along with the rhythms of diurnal and weather cycles, which often overtook the planned rhythms of the event. It was the interference among these rhythms that was strongly articulated over the course of the weekend. The utopian vision of the three-day retreat structured how they heard, listened to, experienced, remixed, and exchanged the sounds. They heard them as sounds of a changing climate, as art commodities, as impositions on their privatized space, or as the sounds of nature, depending on their cultural practices and postures. Within these embodied experiences, the keynote sounds—the bursting of oil rig motors—moved from mere information to the encroachment of their private retreat soundscapes.

One could, however, also read the recording and remixing of specific sounds as a commodification—a colonial settler extraction—of the retreat experience and an abstraction of the concrete human-ranch interactions. Worse, it could contribute to the ongoing homogenizing soundscapes now associated with "nature"—bird songs, rivers flowing, and so on. We hope, however, that the workshop practices also provided participants with the opportunity to explore and interrogate the embodied cultural practices of listening and to experience listening to themselves in relation to an inhabited soundscape. If Lefebvre is right, it is exactly this type of listening—to the silences, to the "murmurous noises"—that destabilizes the hegemonic, though often disordered, timescales of consumption and destruction. It is a listening that is deeply environmental and ecological, and it is the task of soundscape researchers and teachers to cultivate it as a rhetorical, kinship, and political practice. The next chapter introduces a public history audio project that explicitly resists extracting and commodifying the voices and silences of a displaced population in Denver and, instead, seeks to engage participants in an embodied multisensory interaction with the cultural and political processes of displacement, memorializing, urban renewal, and gentrification.

5 Social Justice and Location-Based Sonic Projects

The sonic practices we identified in our classrooms and our collaborative community projects carry both possibilities and limitations for future research and instruction. To summarize, we have thus far emphasized sonic rhetorical practices that 1) create capacity to receive the vibrational and ephemeral effects of sound on bodies and environments; 2) amplify and attend to those rhythms and sounds ignored or marginalized by dominant cultures; 3) record and document those marginalized sounds; and 4) exchange, remix, and archive them with local and networked audiences. While the Re/Call case study analyzes an ecological listening practice that focuses on how the inhabitants of a particular outdoor place influence one another via sound and the cultural meanings indicated by those experiences, the songwriters study moves inside to analyze how resonant listening operates for singer-songwriters within regular collaborative song circles and recording sessions. These collaborative practices of sonic composing—receiving, amplifying, recording, and exchanging—encompass the resonant listening that informs songwriters' recording sessions and song circles as well as the environmental art exhibit. Such practices serve to generate diverse approaches for sonic rhetoric. As we mined these locations for sonic rhetorical practices and pedagogical lessons, we also investigated more deeply our own participation in the musical and artistic experiences.

At the same time, we recognize the limitations of our locations and the selection of participants. While participants in the Re/Call event and song circles posed numerous challenges to dominant art and music cultures, they also curated and selectively monitored human and nonhuman movement across the boundaries of "art," "exhibit," "music," "community," and "nature." Even our basic feminist assumptions and other efforts at inclusivity ultimately re-asserted rhetorical practices that welcomed certain available and privileged people over others. Our reflections on receiving sounds, to name just one sonic practice, can easily presume an able-bodied human listener who simply needs to attune more closely to the sounds in their environment or presume that all participants experience sounds, noises, rhythms, and silences in the same "shared" able-bodied way. In their groundbreaking chapter "Sound and Access: Attuned to Disability in the Writing Classroom," Dev K. Bose, Sean Zdenek, Prairie Markussen, Heidi Wallace, and Angelia Giannone articulate a pernicious and exclusionary tendency in scholarship on sonic rhetoric—the tendency "to inscribe the able-bodied subject in shared experiences of hearing." In our leap to celebrate the communal and collective aspects of sonic composing, we often assume the experience of hearing a particular soundmark or song, and even the experience of "communion" itself, is the same for everyone. Also, when it comes to producing soundscapes or songs, our social and personal identities can quickly coalesce around our training and ability to receive, amplify, record, and remix certain sounds—like someone who can identify a G chord or the song of a migrating cedar waxwing—such that any (inevitable) deviation from that identity is often registered as a loss of social or personal power. As Zdenek notes, "Scholars in disability studies have taken issue with the hierarchy that positions disability only as a terrible absence." He cites H-Dirksen Bauman and Joseph Murray's concept of "deaf gain" as an important epistemological and pedagogical shift from a paradigm of normalcy (e.g., where disability is framed as loss) to a paradigm of rich multisensory diversity. For example, the experience of deafness can bring additional multisensory forms of communication to a rhetorical situation, such as "enhanced and prolonged eye contact, intersubjective engagement, collectivist social patterns, transnational bonds, less auditory distraction, and acute visuospatial aptitudes" (qtd. in Zdenek).

The publication of Bose et al.'s chapter coincided with Staci M. Perryman-Clark's CFP for the 2022 CCCC, entitled "The Promises and Perils of Higher Education: Our Discipline's Commitment to Diversity, Equity, and Linguistic Justice." Like Bose et al., Perryman-Clark foregrounds issues of access to classrooms, universities, and disciplinary conversations and frames them as issues of social justice vs. issues of personal failing or loss. She asks:

> In the spirit of inclusivity, how do we practice diversity in our teaching—I mean, how do we really practice diversity as opposed to simply teaching about it? And how do our practices afford opportunities to both teach and model inclusivity as well as offer spaces to learn from the wide and diverse range of experiences that students bring with them when they enroll in higher education more broadly and in our writing courses more specifically?

Perryman-Clark highlights the current stakes for Black and Brown academics and students for whom social justice is not merely an academic conversation but a life and death issue (e.g., Black and Brown Americans are much more likely to be arrested, incarcerated, and killed by police than white Americans; Black and Brown Americans are much more likely to die or suffer from illness than white Americans; and Black and Brown Americans are much more likely to die from or be displaced by environmental disasters and/or gentrification than white Americans). She then asks us to consider how we are educating students in the pursuit of social justice and to note the perils for not doing so.

In the remaining pages of this book, we want to continue this conversation and inspire other researchers to do so as well. Here, we describe our new and ongoing research that helps to answer the crucial quality of life questions raised by Perryman-Clark and Bose et al. Case studies of convenience or access, like ours, operate within the constraints of our physical, cultural, and institutional limitations and selection processes. Each focused study does capture a distinct angle on a network of sonic practices based on location and thus helps to honor diverse ways of participation when receiving and producing sound. How can our sonic events and locations—our workshops, classrooms, and research sites—acknowledge and celebrate an even more diverse set of multisensory rhetorical practices and experiences? How can we acknowledge and celebrate additional marginalized experiences, while also staying true to the forms and locations of song circles, music studios, soundscape workshops, art exhibits, and other sonic events? At the same time, how might we reframe these experiences within broader global environmental and social justice goals and movements?

Finally, a question even more intimate but difficult to answer is this: how can our written words here in this book (mis)communicate our own culturally informed multisensory and embodied forms of listening? When Krista Ratcliffe first argued for the practice of what she called "rhetorical listening," she applied it to both spoken and written discourse. She defined rhetorical listening as "the performance of a person's conscious choice to assume an open stance in relation to any person, text, or culture" (26). It requires, she argued, a "desire for intersubjective receptivity, not mastery" (28). For Ratcliffe, such receptivity means "letting discourses wash over, through, and around us and then letting them lie there to inform our politics and ethics" (28). "Letting discourses wash over, through, and around us" is more than a metaphor; it is an embodied practice of listening—albeit an idealized one. As we have argued throughout this book, our listening practices are informed by largely unconscious cultural filters, physical processes, and trauma histories. From the beginning, Ratcliffe frames rhetorical listening as an ideal, which will be experienced and practiced differently depending on individual bodies and cultures, and these differences come to surface if one is also listening for them. It is this kind of attention—attention to both the sounds and the ways we, as white middle-class academic cisgendered female researchers, are hearing and processing them—that we seek to cultivate in our sonic projects. Of course, as we have maintained throughout this book, we also want to expand our attention to those sounds that are not discursive or linguistic but still powerfully shape our physical and cognitive reactions and processes. What follows in this chapter are descriptions of two research projects that were inspired by our own model of sonic composition, as well as urgent calls for more diverse forms and models of listening in sonic research.

The 9th Streeter Podcast Project

One way we have begun to address issues of social justice, access, and academic privilege in our research and instruction is through Michelle's advanced composition class's "9th Streeters" Podcast Project. The project expanded the campus soundscape analysis assignment detailed in chapter two to address questions of gentrification, social justice, and linguistic diversity. Often students in Michelle's classes respond to the campus soundscape analysis assignment by recording sounds from a historical park located on the south side of campus—9th Street Park—which includes two rows of historical homes and a green space preserved from the once vibrant Chicana/o neighborhood displaced by the city and the campus in the early 1970s. In 1972, an entire Chicano/a neighborhood was displaced by the Housing and Urban Development Agency and the city of Denver in order to build the Auraria Higher Education Center (AHEC) campus. Residents organized the Auraria Residents' Organization (ARO) to resist the displacement, but ultimately voters supported the plan to clear Denver's oldest neighborhood and build a modern campus that now contains Community College of Denver, Metro State University, and the University of Colorado Denver. The displacement was just one example of urban renewal projects that were happening at the time across the nation, where HUD would deem whole neighborhoods "blighted," so they could seize real estate for development. Part of AHEC's deal with the Auraria neighborhood was to provide a "Displaced Aurarian Scholarship" for families and their descendants, which is where the name "Displaced Aurarian" came from. Before then, residents called themselves "Westsiders" or "9th Streeters," if they or their families lived on the historic park. Last fall Michelle's students composed a podcast in order to connect digital content and oral stories from those Westsiders affected by the displacement with the current physical architecture and landscapes of 9th Street Park, thus creating a hybrid (digital and physical) experience for participants. The project centered on one central question: what historical and cultural voices and sounds have been excluded and/or policed in the gentrification of the campus soundscape?

Besides engaging students in the rhetorical and technical aspects of producing an actual podcast (the final podcast was an episode of the grant-funded Department of English's podcast *CultureKlatsch*), the project asked them to contribute soundscapes and other sonic material to an already rich archive on the *Displaced Aurarians* and, more particularly, the 9th Streeters. Although students studied the larger Auraria neighborhood and AHEC, an administrative body that continues to manage and maintain the buildings on what is now called the Auraria campus, 9th Street was of particular interest not only because the public can tour the houses and learn about those who lived there, but also because the majority of English faculty and department offices, including Michelle's, are located in five of the houses. In a recent online symposium featuring three 9th Streeters, Michelle heard them mention feeling unwelcome at the houses now that they were department and university offices, and she wanted to learn more about their discomfort, especially since she spends so many hours each week there. The course project addressed the question of what it would take, if anything, for 9th Streeters to feel welcome back on campus. What kinds of reparations (monetary and otherwise) need to be made by the department and university to honor the individuals and community that once thrived there and make the campus and its resources more accessible to them?

All three universities located on the Auraria campus have expanded the availability of scholarships to the descendants of Displaced Aurarians, but is that enough? Many have stated that accessing information about the scholarships is difficult, even with the help of advocates in the Auraria library. Students in Michelle's podcasting class collected sonic artifacts and interviews regarding these scholarships, reparations, and life on 9th Street Park before and after the displacement. The podcast also made explicit connections among regional gentrification, transportation, and fossil fuel industry practices that have affected Displaced Aurarians since the 1970s. Some community members, for example, have had to move a second time, due to ongoing development and highway expansion in the Denver metro area.

In many ways this project extended research in what Ahern calls the growing subfield of soundscape studies ("Soundscapes"). Ahern highlights the multi-dimensional aspects (sound sources, sounds unfolding through time, sounds occurring simultaneously, and sounds moving within and across space) of soundscape design and composition, which makes it especially promising for rhetorical studies, a field long concerned with situated approaches to communication. Ceraso similarly emphasizes the importance of soundscape design and analysis to situated multimodal listening pedagogy. She argues that "Schafer's work on soundscapes is especially relevant to multimodal listening pedagogy because it underscores the ecological relationship among sound, bodies, and environments as an essential part of the process of listening to and composing soundscapes" (70). "According to Schafer," she writes, "sonic composing practices must account for how humans experience, interact with, and contribute to the soundscapes they inhabit" (70–71). Ceraso rightly expands on this aspect of Schafer's framework, stressing the "entire ecological system of which sound is a part" (71). It is a system that includes the materials of architecture, furniture, bodies still and in motion, and technologies—digital and otherwise (73).

Our own research on soundscape analysis and assignments has emphasized the cultural factors that produce specific soundscapes and the cultural conditioning that guides our attention toward certain sounds over others. The campus soundscape assignment outlined in chapter two, for example, asks students to link sounds and sound walls to cultural and institutional values, and many students on Michelle's campus have made obvious connections between the sounds of traffic, buses, and light rail trains with the university's identity as an urban campus. The 9th Streeters podcast project asked Michelle's students to take a more critical look at their campus' history of gentrification and urban renewal and highlight how the campus soundscapes (including its voices, ambient sounds, sound walls and barriers, etc.) are informed by this cultural history. Michelle also asked her students to consider the policing practices (sonic and otherwise) that inform the boundaries of the urban/community corridors on campus and the surrounding neighborhoods. How does the campus police construct its brand and its boundaries with neighboring communities? As abolitionist and community organizer Mariame Kaba argues, it is usually those who are the least resourced—in this case, the residents in the surrounding Black and Brown neighborhoods—who are subjected to "everyday mundane surveillance" (139) whether it is by actual police officers or members of groups who "belong" there.

Ahern claims the main rhetorical purpose of soundscape design is to "engage the listener in a particular sense of space, as well as the experience of inhabiting and navigating that space, regardless of whether that space is virtual or physical, regardless of the experience of the soundscape—visually, sonically, kinesthetically." In order to construct a multisensory experience for listeners, Michelle's students conducted interviews with two 9th Streeters, asking them to describe in somatic detail what it was like to inhabit and navigate the neighborhood before and after displacement. At the same time, her students engaged in a series of soundwalks, where they received, amplified, and recorded a diversity of sounds (voices, noises, silences, etc.) in a number of gentrified (or gentrifying) locations. In her own soundwalking assignment, Ceraso asks her students to engage in multimodal listening as they walk together silently through various parts of her campus. Because it is difficult to recall such ephemeral experiences after the fact, Ceraso asks students to pause every fifteen minutes and record their experiences and interactions with the campus soundscape *in situ*, then share them with each other back in the classroom. After reflecting for a while on their soundwalks, Ceraso's students identified a contradiction between the university's brand as an urban university and its rural bucolic soundscape. Such contradictions also arose in Michelle's course. The sounds and narratives of gentrification arose in contradiction to the branding sounds and narratives of the campus and university, as well as the region. To make such contradictions more explicit from the beginning, the students who produced the final podcast voiced their own ambivalence about the gentrification process. As graduate students in English Studies, they largely benefited from gentrification processes that made art and multiculturalism more accessible. Teaching at a Hispanic-serving university, for example, would give them some economic capital on the job market, while learning about the history and culture of the Displaced Aurarians would give them a fair amount of academic capital, as they used what they learned to complete Master of Arts requirements and acquire additional knowledge and skills. However, over the course of the project, they began to see how they, as well as the Displaced Aurarians, were affected by the contradictions between the university's mission and its gentrification practices.

Before they embarked on their own gentrification research, Michelle's students studied the creation of soundscapes, sonic archives, and sound maps in other cities, many of which record the impact of gentrification, development, and language policies on communities and ecosystems. In her own sound map assignment discussion, Ceraso explains the general rhetorical purposes of sound maps, a relatively new sonic genre that depends upon GPS and/or web technologies: "While different sound maps have different purposes, generally speaking sound mapping is intended to give users an opportunity to listen to and examine the soundscape of a place to get a better sense of what life is like there; to document changes in the soundscape over time; and, for people who already inhabit those spaces, to amplify sounds that they may not notice anymore due to overexposure" (93). Besides addressing which sounds are amplified and which sounds are silenced, Michelle's students examined who benefited from certain sound maps, especially the more popular ones linked to tourism in cities like New Orleans, New York, and Chicago. Are these maps further exploiting communities that have already carried the costs of gentrification? And, if so, how might they better contribute to these communities? "Cultural piracy" was a key term in our analysis. Urbanist and historian Devarian Baldwin emphasizes this concept in his recent book on universities and gentrification. He argues that when, for example, the University of Chicago decided to move a popular blues bar to its campus-owned shopping district instead of resourcing the Bronzeville neighborhood that originally housed the bar, restoration advocates accused the university of "cultural piracy" (2). The Auraria campus made similar moves when they decided to retain and move several of the Victorian homes and the Spanish style church, St. Cajetan's, in order to repurpose them as charming campus offices, classrooms, and labs and, at the same time, protect them by placing them on the Colorado Registry of Historical Properties. Michelle's students asked how this move benefitted and/or harmed not only the original owners but also those living in adjacent lower-income neighborhoods.

Baldwin's book, *In the Shadow of the Ivory Tower*, traces the economic and political history of urban universities and their effects on surrounding neighborhoods. It is a history that resembles AHEC's own development projects beginning in the 1970s. He writes:

> When most of the United States had abandoned cities in the mid-twentieth century, higher education was one of the only institutions that remained. A core group of colleges and universities used public urban renewal money to bunker themselves behind the walls of campus buildings or demolished city blocks—and away from the growing 'invasion' of Black and Latinx residents. But starting in the 1990s, young professionals, empty nesters, and the children of suburban sprawl began to seek a more urbane lifestyle. And municipal politicians and real estate developers from different cities started competing with one another to capture the potentially lucrative new tax base and its consumer dollars. At the same time, colleges and universities were looking for new revenue streams in the face of tight state budgets. The interests of university and city leaders converged when the college campus was reimaged as the palatable and profitable version of a safe urban experience. (7)

Baldwin argues that many urban schools, like AHEC, "have become the dominant employers, real estate holders, health-care providers, and even policing agents in major cities around the country. And the lower-income neighborhoods and communities of color that stand in the immediate path of campus expansion, while in deep need of new investments, are left the most vulnerable" (6). Much of this expansion, according to Baldwin, relies on the wider cultural assumption that "higher education is an inherent public good" (12). Though this belief has been challenged by conservative politicians in recent decades, who continue to defund public higher education at the state and federal levels, universities still rely on it for pursuing and sheltering "their own interests in for-profit research or even the financial security of private developers and investors that sit on their campus land" (Baldwin 12). While Michelle's students did not ignore the positive impact AHEC has had on the community (indeed, many of those displaced celebrate the impact the scholarships have had on their families), they did examine who in the community and city has paid the cost for the campus' land acquisition, growth, and maintenance, and they ultimately focused on those residents who have historically fought against inequitable community development and have pursued alternative university-community partnerships, including increased accessibility to university resources and reparations.

Finally, Michelle's students studied the processes and effects of transcription. The Denver Public Library houses hours of oral interviews with Displaced Aurarians, and students in the university's public history program have begun transcribing many of them. The transcription to print will allow multiple audiences to access them digitally beyond the archive's library walls. Because transcription is a common form of remediation for radio and sonic archives, Michelle's students studied the pragmatics and politics of the practice and what it affords listeners and readers. What does reading the transcript offer readers that is different from the sound of voices and noises on the tape? And what kinds of struggles do the transcribers face when documenting dialects and non-discursive content like laughs and sighs? Instead of framing the differences between the two texts as gains and losses or advantages and disadvantages, the students explored how the histories—the written and the oral—allowed for different experiences and processes of communication and meaning making, and they researched how these transformations played out in their own interview transcriptions, as well as in the interviews they accessed during their research.

The Final 9th Streeter Podcast Episode

Based on their research on the Displaced Aurarians, three of Michelle's students—Katherine Allen, Joshua Voiles, and Natalie Waddington—ultimately chose to co-produce and co-publish the 9th Streeter episode for the department's Fall 2021 *CultureKlatsch* episode. Before beginning their research, they read and discussed Eric Detweiler's article on the multi-genre and extra-discursive aspects of podcasts. Detweiler argues, "If we and our students are to realize the rhetorical and pedagogical possibilities of podcasts, to engage their potential as an activity 'around which writing and communication can be organized,' we should not reduce the medium to the essayistic manifestations that may feel most familiar to us." "Reducing the wide-ranging world of podcasts to one well-trodden form," he continues, "tunes out the inventive possibilities—the fresh genres, modes of inquiry, and media combinations—that are key parts of what podcasting offers." Following Detweiler's invitation, Katherine, Josh, and Natalie explored the rich sonic resources of on-location recording, music, and a collaborative inquiry-based approach to the Aurarian displacement and the city's history of gentrification. When one listens to the published episode, one hears a series of questions and interview footage with a great deal of ambient noise and varying volume levels. This was the first time these students had produced a podcast, and they wanted to do it collaboratively as what Detweiler would call a "discussion podcast," without a singular host or single narrative arc. Instead, they allowed the podcast to unfold from their own ambivalent cultural beliefs and values around gentrification and their own cultural and economic investments in it. You can hear the entire final podcast, entitled "Gentrification: The Denver Experience," here: www.cultureklatsch.com.

Listeners to past *CultureKlatsch* episodes will hear stark differences in Allen et al.'s approach. The episode is filled with ambient noise and sudden shifts in sound levels and recording quality, as it moves from studio, to outdoor, to office, then to coffee shop locations. The sudden shifts in location draws attention from the content of the speech to the locations themselves. Aesthetically, making the whole podcast "noisy" with ambient sounds may have been the better approach; however, the sudden shifts remind listeners that they are in a different location with different people. Not only do we hear changes in what Roland Barthes called the "materiality" of the "grain of the voice" (506), but also changes in the grain of the location and the reverberation of bodies and objects in those locations. In one sense, the abrupt shifts mirror the disorienting processes of displacement and gentrification, though without the extreme economic, social, and physical effects, of course. Depending on one's cultural and audio training, these shifts will be heard as richly meaningful communication signals and/or meaningless annoying static. In chapter two, we discussed how important Hammer's definition and theorization of sonic noise is to our pedagogy, and his points are relevant here, too. As Hammer argues, there is no such thing as noiseless communication. Instead, it is embedded within every communication event and system, and whether we hear a sound as noise or signal is arbitrary—an effect of our cultural and political experiences and training. If we do hear a sound as noise, we may choose to either work harder to discern the message or simply ignore it. In relation to the 9th Streeter podcast episode, listeners, who are often already distracted, may decide "making sense" of its messages across numerous shifts in tone, cadence, and volume is not worth the labor.

For those who do commit to listening through the various, sometimes startling, shifts in location, volume, and microphone quality or for those who commit to reading the transcript, the episode asks a variety of stakeholders (9th Streeters, experts on urban planning and housing, campus community members) about the economic, political, physical, and emotional costs of displacement and gentrification.

The podcast begins with Natalie introducing all three hosts then quickly segues into Josh's reflections on his life as a student on the Auraria campus: "The campus has basically become a second home." All three hosts laugh about how they are now producing the podcast from the very campus—more particularly, the library's podcast booth—which they consider an integral part of their everyday lives. "It's my office, it's my library, it's my gym. It's where I get the majority of my caffeine," Josh continues. "But it's not always been the case." Josh then transitions to a more historical perspective: "Before the executive plans and the bulldozers, Auraria was Denver's oldest community." Josh's own everyday practices and investments in the campus become both the context and impetus for grappling with its oppressive history and frames the first interview, which Katherine conducts with two Auraria campus employees: Francine Olivas-Zarate, a program coordinator on campus, and her sister Benita Olivas, an advisor. Their family members lived on 9th Street Park before the mass displacement, and both have worked on campus near their family's original home.

The ambient traffic noise in the interview segment tells us we are outside where Benita and Francine's voices compete with a variety of ambient "noises." As they tell their stories, we may imagine what the neighborhood used to sound like and all the ways the sounds of displacement, development, and commuter campus life continue to compete with the voices of past and present inhabitants. "Everybody would take care of your kids. Everybody was there. They had people who played guitars and sang and danced. They would get together on certain evenings here and just entertain," Benita began. "They did a lot of that in this house right here," Francine adds. In retrospect, the podcasters could have done more to contrast the pre- and post-displacement soundscapes. What cultural, community, and environmental sounds have been lost because of the displacement, bulldozing, campus development, and preservation? This question is being taken up by Michelle's current students, as part of a larger audio tour project.

Instead of composing contrasting soundscapes, the *Cultureklatsch* podcasters rely on visual descriptions to represent the dramatic changes brought about by urban renewal and gentrification in the city. Natalie describes what downtown Denver looked like in 1976 followed by Katie's description of what Auraria looked like before the displacement, as well as the racist zoning policies and politics that shaped its development. The podcast moves from the remembered sounds of neighbors dancing and singing together to visual representations of the houses, churches, and businesses that once comprised the neighborhood to the more widespread racist zoning practices. The script does not emulate the strong narrative arc common to popular podcasts like *This American Life* and *Radiolab*; instead, it takes us on a trip—sometimes with rapid turns—through a layered oral and visual archive of displacement and gentrification. Not every listener will be willing or able to take such a trip. The podcast continues to shift locations through time and space as we learn about the flood that devastated the neighborhood and made it vulnerable for redevelopment by the city, then listen to Josh's interview with geography and urban planning scholars Brian Page and Jeremy Nemeth about the history and politics of zoning in the city. We learn, from another shift in location, that the neighborhood's property values were driven lower by the flood, racism, and its proximity to the city's growing manufacturing district, and it was those lower property values that made it attractive to the campus developers and the Denver Urban Renewal Authority.

After a brief interlude into a discussion of the Displaced Aurarian Scholarship, the podcast suddenly moves to another gentrification hot spot—the Five Points neighborhood and the redeveloped RINO district, where Natalie lives. We hear, again, about how these gentrified areas are an integral part of these students' lives—a pleasurable source of eating, drinking, dancing, and socializing (pleasures once experienced within the Displaced Aurarian community). Natalie then transitions to the costs of such development and how the rising house prices have created a crisis in the city. Her interview with Laura McKenna, Development Director of HomeAid America, focuses the podcast on this growing housing crisis.

Finally, the podcast jumps to a large-scale discussion of the multiple displacements and intergenerational trauma Aurarians have faced since the 1970s, due to ongoing gentrification, housing markets, and political decision makers. The larger cycle, according to Nemeth, is disinvestment in a place (e.g., suburbanization), redlining, decline, and then redevelopment and displacement into more vulnerable neighborhoods. "It's this cycle of investment and disinvestment," he argues, "and it is not a mistake." Ultimately, the hosts land on two political solutions for displacement and gentrification: a recommendation by McKenna that housing in Denver become income-based and a recommendation by Nemeth that we maintain and increase publicly subsidized housing. Again, all these shifts in location, scale, volume, tone, and theme can disorient the listener, but the changes also signal multiple layers of a rich archive of gentrification. Instead of one monolingual narrative, we experience multiple nodes, voices, noises, and silences within a network of city and private development practices that weaponize racism and natural disasters for profit.

The final "9th Streeter" episode has since become part of a larger campus-wide project to collect archival and pedagogical materials on the history and politics of the displacement. The larger project also includes oral history workshops, symposia, the rededication of homes along 9th Street Park to the Displaced Aurarians, an extension of the Displaced Aurarian Scholarship, and a grant-funded three-year effort to memorialize the displacement in the curriculum, community, and campus. A big portion of the latter memorialization is a digitized soundwalk (audio tour) of the campus, which includes portions of Michelle's students' podcast and stories recorded during the oral history workshops. What Michelle had hoped would happen with the podcast—a "walking" soundscape/oral history of 9th Street Park and other cultural landmarks on campus and in the neighborhood—is becoming part of this interdisciplinary community-based project over the next several years. Like Ceraso, Michelle is concerned about the ableism inherent in the term "soundwalk." The term implies that only a walking body can experience the sounds and movement of sounds within a soundscape as they are *supposed* to be experienced. Such an assumption belies the diverse experiences of students, staff, and faculty as they navigate campus buildings and classrooms physically and digitally via differently abled bodies, vehicles, and interfaces. Ahern notes that "even among students with relatively similar physiological listening abilities, there should not be an assumption that all participants, designers, or composers of a soundscape experience listening or draw on listening practices in the same way." As the Auraria audio tour project develops over the next few years, researchers and composers are explicitly taking up questions of ableism and access in relation to the podcast and sonic artifacts they are collecting, but they are also addressing them in relation to the physical architecture of 9th Street Park. Because the offices on the park are part of the historical registry, the campus does not have to provide ramp and/or elevator access to students. Thus, the offices are actually not accessible to many of our students and perhaps not accessible to particular 9th Streeters who may want to return to visit their former homes.

The research is also focusing on accessibility in relation to language use by exploring the connections between language use and social and institutional privilege on campus and in adjacent neighborhoods. Historically, Spanish has been spoken on 9th Street Park—by the displaced community members, as well as the staff, students, and faculty members who have worked and studied on the park. The language has been symbolically honored by the Chicano/a heritage centers located on the park and the university's new status as a Hispanic-serving institution. One of the main questions for Michelle's current students is how language has informed access to resources on campus and in the neighborhood.

Displaced Aurarians Memory Project

In a 2021 issue of *Kairos*, Valenzuela and Landa-Posas published a presentation on their work with mobile recording technology, audio narratives, and space. For their presentation, they conduct their own soundwalk as they document Landa-Posas' memories along a well-known boulevard in her hometown, which was undergoing radical redevelopment and gentrification. They frame their soundwalk as an intervention in the larger silencing of multilingual immigrant communities and within the Chicana/Latina feminist methods of *testimonio*, which "centers the subaltern lives . . . allows them to narrate their everyday experience, [and] allows the subaltern to theorize from their own bodies." They also engage what Margaret Ramirez calls a "borderland analytic," which "offers spatial attention to how power relations emerge and are contested in gentrifying spaces" (15). Thirdly, they theorize their listening bodies through Gloria Anzaldua's borderland theory, which employs an imaginative political practice of tuning into "material forms of liminality" and from there "our radical interconnectedness." For Anzaldua, this interconnectedness includes the "discursive human and nonhuman, the corporeal, and the technological." Valenzuela and Landa-Posas use Anzaldua's "liminal imagining" practice "to render audible, or more so, to sonically and viscerally *testimoniar* alongside the sounds that exist within and on the margins." Thus, when producing their soundwalk, they first collected ambient sounds from their walk down a familiar street, then recorded their narratives and *dichos* (spoken sayings) while driving across the borders of several neighborhoods in order to highlight the effects of gentrification. The sounds they recorded along the way became a "third voice with and in between us, alongside us, and in front of us." As they engaged with them and reflected on them together, the sounds afforded new sonic relationships; and, as they produced the soundwalk and shared it with the conference and *Kairos* audiences, a growing sense of agency and interconnectedness emerged.

The audio tour Michelle is producing for the Displaced Aurarian Memory project employs many of the same Chicana/Latina feminist methods, with the focus on sounds as the facilitators of the imagined past and future and on the audio tour as an intervention in the silences created by gentrification on campus and in the surrounding neighborhoods. The oral history workshops and the rededication of homes, which have already begun, have focused on recreating a sense of relationship among Displaced Aurarians and their neighborhood and community. Many of them have been displaced multiple times within the cycles of disinvestment and reinvestment and have difficulty even imagining, let alone feeling, a sense of community or interconnectedness.

At the first Displaced Aurarian oral history workshop Michelle attended, sponsored by History Colorado, participants were asked to first draw maps of their neighborhood, then to write and share the stories that came to them. Michelle noticed from her seat along the wall that the oral stories had already begun, as participants started whispering shared memories with each other. Later, it came out that generations of each family were sitting together at tables, which facilitated this kind of intimate dialogue. Also, all the workshops took place in St. Cajetan's Church. Now an event center and classroom space for campus, the desacralized church had been an epicenter for community life in the Westside neighborhood. The priest, Father Pete Garcia, had been an outspoken community and political organizer against the displacement, and his efforts were ultimately not supported by the larger diocese. It was understood in the community that this betrayal largely facilitated the displacement. Naturally, many of the stories shared with the workshop audience were about activities in the church. The visual maps were less compelling than the space of the room itself. When the workshop facilitator asked participants to share their stories via microphone, only one read from any written narrative. Instead, they had gathered a collective narrative from the table, with family members and neighbors supplying supporting and corrective details. Those formative dialogues were unfortunately lost in the ether as they shaped the center stage monologues at the microphone.

Despite this loss of important dialogue, many of the shared narratives are becoming part of the soundwalk/audio tour, as they jolt listeners out of their everyday visual sense of the campus to tell them about the factories, restaurants, homes, and, most importantly, relationships that once formed there and were silenced by the radical displacement. Some of the narratives include what participants called "ghost stories." As they walked the campus and sat in the church, several participants were visited by images of relatives who had passed, telling them something important about their deaths and illnesses. Such visitations and memories allow participants to vocalize narratives of death and illness caused by oppression, overwork, and lack of resources. These "ghost stories" can be considered what Anzaldua calls "liminal imaginings" or what Valenzeula and Landa-Posas call the practice of rendering audible subaltern voices and "to sonically and viscerally *testimoniar* alongside the sounds that exist within and on the margins." Framing the Displaced Aurarian soundwalk as *testimonio*, as a "borderland analytic," is important in order to challenge potential cultural appropriations and what Baldwin calls "cultural piracy," which may already be in play with the well-publicized rededication ceremonies of certain homes along 9th Street Park. Questions over who has access to, who "owns," and/or who benefits from the soundwalk and the larger efforts at memorialization are being foregrounded and included as content, alongside the current voices and sounds of ongoing gentrification on campus and in the surrounding neighborhoods.

Futures in Sonic Rhetoric

In the end, we wrote *Composing Resonance: Culture and Collaboration in Sonic Practices* because we want to encourage and collaborate with other researchers on projects that make social and linguistic justice central to sonic rhetoric in classrooms more generally. As one example, Mack Curry IV, who worked with Mary as a graduate student, designs and studies linguistically diverse multimodal pedagogy for his English composition courses to offer a type of linguistic social justice. In "Adopting Home Language and Multimodality in Composition Courses," Curry emphasizes many of the same sonic assignments as other researchers in this book, but with explicit goals for student code-meshing with rhetorical awareness. By examining his own classroom's space, Curry answers Perry-Clark's calls for privileging and expanding linguistic diversity in all our research and teaching. We hope future research in sonic rhetoric begins with questions of linguistic, social, and economic justice and that it privileges the diversity of voices, sounds, and experiences within any performance, text, or soundscape. We also hope that researchers join public historians and activists to foreground the voices of those most marginalized and vulnerable and work to provide access to resources for these populations.

We recognize our own responsibility to engage continually in resonant listening practices—the receiving, amplifying, recording, and exchanging—that create conditions for diversity, equity, empathy, collaboration, and care in our universities, communities, and environments. At the same time, we recognize the importance of continuously questioning and even un-doing the cultural norms that inform those listening practices. Both reduced listening techniques and cultural soundscape analyses can assist in this ongoing reflective and transformative somatic process. However, we need to apply these techniques in increasingly diverse locations with all kinds of people as they create moments of connection through sonic production practices. As our case studies have shown, sonic enculturation is an incomplete, limited, and location-based process that changes with our everyday encounters with material conditions. This inherent incompleteness makes any sonic process vulnerable to critique and transformation. When we begin noticing how our bodies and environments receive and amplify particular sonic vibrations over others, we begin to witness how these sensations shape our very sense of self and other and body and environment. Here the teachable moments begin. These experiences can perpetually inform our sonic recording and exchange practices and move us into a deeper, more inter-relational understanding of human and nonhuman audiences.

Works Cited

Ahern, Kati Fargo. "Soundscapes: Rhetorical Entwinements for Composing Sound in Four Dimensions." *Tuning in to Soundwriting*, edited by Kyle D. Stedman, Courtney S. Danforth, and Michael J. Faris, Intermezzo, 2021, intermezzo.enculturation.net/14-stedman-et-al/ahern-4.html.

—. *The Sounds of Rhetoric, The Rhetoric of Sound: Listening and Composing the Auditory Realm*. 2012. North Carolina State U, PhD dissertation.

Alaimo, Stacy. *Bodily Natures: Science, Environment, and the Material Self.* Indiana UP, 2010.

Alexander, Jonathan. "Glenn Gould and the Rhetorics of Sound," *Computers and Composition*, vol. 37, no. 3, 2015, pp. 73–89.

Alexander, Jonathan and Jacqueline Rhodes. *On Multimodality: New Media in Composition Studies*. CCCC/NCTE, 2014.

Ayers, Andrew. "Sick Architecture and the Spatial Politics of Disease." *Pin-Up*, 2022, pinupmagazine.org/articles/sick-architecture-exhibit-brussels-beatriz-colomina.

Baldwin, Davarian L. *In the Shadow of the Ivory Tower: How Universities Are Plundering Our Cities*. Bold Type Press, 2021.

Ball, Cheryl, and Byron Hawk, eds. *Sound In/As Compositional Space: A Next Step in Multiliteracies*, special issue of *Computers and Composition*, vol. 23, no. 3, 2006.

Barbee, Libby, and Bill Nelson. *The Sound Mirror Project*. 2018, RedLine, Rocky Mountain\ Land Library.

Barthes, Roland. "The Grain of the Voice." *The Sound Studies Reader*, edited by Jonathan Sterne, London: Routledge, 2012, pp. 504–10.

Batiste, Jon. "Building Community Through Music." *Chicago Ideas*, 30 Jan. 2015, *YouTube*, youtube.com/watch?v=d5n2DwjrfMY.

Berlant, Lauren. *Cruel Optimism*. Duke UP, 2011.

Blauert, Jens. *Spatial Hearing*. MIT P, 1999.

Bose, Dev K., et al. "Sound and Access: Attuned to Disability in the Writing Classroom." *Tuning in to Soundwriting*, edited by Kyle D. Stedman, Courtney S. Danforth, and Michael J. Faris, Intermezzo, 2021, intermezzo.enculturation.net/14-stedman-et-al/bose-2.html.

Bowie, Jennifer. "Podcasting in a Writing Class? Considering the Possibilities." *Kairos: A Journal of Rhetoric, Technology, and Pedagogy*, vol. 16, no. 2, 2012, kairos.technorhetoric.net/16.2/praxis/bowie/index.html.

Breene, Samuel. "The Instrumental Body in the Age of Mozart: Science, Aesthetics, and Performances of the Self." *Early Music*, vol. 42, no. 2, 2014, pp. 231–47.

Britton, James, "Notes on a Working Hypothesis." *The National Writing Project Network Newsletter*, vol. 1, no. 3, 1979, pp. 1, 9–11.

Burgund, Halsey. *From Here to Where*. 2018, RedLine, Rocky Mountain Land Library.

Campbell, Karlyn Kohrs. "The Ontological Foundations of Rhetorical Theory." *Philosophy and Rhetoric*, vol. 3, no. 2, 1970, pp. 97–108.Cardiff, Janet. "Wanås Walk." 1998. *Janet Cardiff & George Bures Miller–Walks*, cardiffmiller.com/walks/wanas-walk/.

Caulkins, Mary, and Louise Martorano. "Re/Call Invitation for Proposals," E-mail to the author. 12 May 2017.

Ceraso, Steph. "(Re)Educating the Senses: Multimodal Listening, Bodily Learning, and the Composition of Sonic Experiences." *College English*, vol. 77, no. 2, 2014, pp. 102–23.

—. "Sonic Scenes of Writing." *College English*, vol. 84, no 4, March 2022, pp. 311–34.

—. *Sounding Composition: Multimodal Pedagogies for Embodied Listening*. U of Pittsburgh P, 2018.

—. *Sounding Composition, Composing Sound: Multimodal Listening, Bodily Pedagogies, and Everyday Experience*. 2013. U of Pittsburgh, PhD dissertation.

Chion, Michel. "*La Rond*." 1983. *YouTube*, youtube.com/watch?v=hLGaamrPQ7E.

—. "The Three Listening Modes." *The Sound Studies Reader*, edited by Jonathan Sterne Routledge, 2012, pp. 48–53.

The City. Directed by Ralph Steiner and Willard Van Dyke, American Institute of Planners, 1939.

Clark, Gregory. *Civic Jazz: American Music and Kenneth Burke on the Art of Getting Along*. U of Chicago P, 2015.

Clifford, James. *The Predicament of Culture: Twentieth-Century Ethnography, Literature, and Art*. Harvard UP, 1988.

Cobos, Casie, Gabriela Raquel Ríos, Donnie Johnson Sackey, Jennifer Sano-Franchini, and Angela M. Haas. "Interfacing Cultural Rhetorics: A History and a Call." *Rhetoric Review*, vol. 37, no. 2, 2018, pp. 139–54.

Comstock, Michelle, and Mary E. Hocks. "The Sounds of Climate Change: Sonic Rhetoric in the Anthropocene, the Age of Human Impact." *Rhetoric Review*, vol. 35, no. 2, 2016, pp. 165–75.

—. "Voice in the Cultural Soundscape: Sonic Literacy in Composition Studies." *Computers and Composition Online,* vol. 23, no. 3, 2006, www2.bgsu.edu/departments /english/cconline/comstock hocks/index.htm.

Cooper, Marilyn. "Rhetorical Agency as Emergent and Enacted." *College Composition and Communication*, vol. 62, no. 3, 2011, pp. 420–49.

Curry, Mack IV. "Adopting Home Language and Multimodality in Composition Courses." *Culture in Focus,* vol. 2, no. 1, 2019, pp. 70–96.

Danforth, Courtney S., et al., eds. *Soundwriting Pedagogies*. Computers and Composition Digital Press, 2018, ccdigitalpress.org/book/soundwriting/index.html.

Davis, Diane. *Inessential Solidarity: Rhetoric and Foreigner Relations*. U of Pittsburgh P, 2010.

Davis, Diane, ed. "Writing with Sound." Special issue of *Currents in Electronic Literacy*, no.14, 2011, currents.dwrl.utexas.edu/2011.

Detweiler, Eric. "The Bandwidth of Podcasting." *Tuning in to Soundwriting*, edited by Kyle D.

Stedman, Courtney S. Danforth, and Michael J. Faris, Intermezzo, 2021, intermezzo.enculturation.net/14-stedman-et-al/detweiler-4.html.

Demers, Joanna. *Listening Through the Noise: The Aesthetics of Experimental Electronic Music*. Oxford UP, 2010.

Eidsheim, Nina Sun. *Sensing Sound: Singing and Listening as Vibrational Practice*. Duke UP, 2015.

Elbow, Peter. *Vernacular Eloquence: What Speech Can Bring to Writing*. Oxford UP, 2012.

Freese, J. H., translator. *Art of Rhetoric* by Aristotle, Harvard UP, 1926.

Frith, Jordan. "Writing Space: Examining the Potential of Location-Based Composition." *Computers and Composition*, vol. 37, 2015, pp. 44–54.

Gentles Stephen J., and Silvia L. Vilches. "Calling for a Shared Understanding of Sampling Terminology in Qualitative Research: Proposed Clarifications Derived From Critical Analysis of a Methods Overview by McCrae and Purssell." *International Journal of Qualitative Methods*, vol. 16, no. 1, 2017, doi.org/10.1177/1609406917725678.

Gewin, Billy. Scheduled Recording Session with Mary Hocks, Cave Cricket Studio, Atlanta, GA, 20 Oct. 2018. http://www.billygewin.com/.

___. Scheduled Recording Session with Charlie Bright, Cave Cricket Studio, Atlanta, GA, 21 July 2019.

Goodman, Steve. *Sonic Warfare: Sound, Affect, and the Ecology of Fear*. MIT P, 2010.

Gries, Laurie E. *Still Life with Rhetoric: A New Materialist Approach for Visual Rhetorics*. Utah State UP, 2015.

Gries, Laurie E. and Collin Brooke. "Introduction: Circulation as an Emerging Threshold Concept." *Circulation, Writing, and Rhetoric*, edited by Laurie E. Gries and Collin Clifford Brooke, Utah State UP, 2018, pp. 3–24.

Gurantz, Maya. "Christmas in 3-D." Narrated by Gurantz. *This American Life*, hosted by Ira Glass, 21 Dec. 2012. *This American Life Radio Archive*, thisamericanlife.org/radio-archives/episode/482/lights-camera-christmas?act=1.

Halbritter, Bump. *Mics, Cameras, Symbolic action: Audio-visual Rhetoric for Writing Teachers*. Parlor Press, 2013.

—. *Sound Arguments: Aural Rhetoric in Multimedia Composition*. 2004. U of North Carolina, PhD dissertation.

—. "Who Wrote That Song?: Writing, Performing, Editing, and Recording." Presentation for "Uncommon Places: The Recording Studio as Compositional Space" Panel, *Conference on College Composition and Communication*, 10 April 2021.

Hall, Alan. "William Thompson IV's War." Narrated by Hall. *Falling Tree Productions*, 21 Dec. 2006, fallingtree.co.uk/?s=William+Thompson+IV %E2%80%99s+War&post.

Hall, Nathan. *Chime Walk*. RedLine, Rocky Mountain Land Library, 2018.

Hammer, Steven R. "Writing Dirt, Teaching Noise." *Soundwriting Pedagogies*, edited by Courtney S. Danforth, Kyle D. Stedman, and Michael J. Faris, Computers and Composition Digital Press, 2018, ccdigitalpress.org/book/soundwriting/hammer/two.html.

Hawk, Byron. *Resounding the Rhetorical: Composition as a Quasi-Object*. U of Pittsburgh P, 2018.

Hocks, Mary E. and Michelle Comstock. "Composing for Sound: Sonic Rhetoric as Resonance," *Computers and Composition*, vol. 43, 2017, pp. 135–46.

Kaba, Mariame. *We Do This 'Til We Free Us: Abolitionist Organizing and Transforming Justice*. Haymarket Books, 2021.

Kahana, Jonathan. *Intelligence Work: The Politics of American Documentary*. Columbia UP, 2008.

Kahn, Seth. "Putting Ethnographic Writing in Context." *Writing Spaces: Readings on Writing, Vol. 2*, edited by Charles Lowe and Pavel Zemliansky, Parlor Press, 2011, pp. 175–92.

Kotz S.A., et al. "The Evolution of Rhythm Processing." *Trends in Cognitive Science*, vol. 22, no. 10, 2018, pp. 896–910.

Lane, Cathy. "Why Not Our Voices?" *Women and Music: A Journal of Gender and Culture*, vol. 20, 2016, pp. 96–110.

The Late Show with Stephen Colbert. CBS. Season 7, Episode 91, 21 Feb. 2022.

Lefebvre, Henri. *Rhythmanalysis: Space, Time, and Everyday Life.* Continuum, 2004.

Livneh, Uri and Anthony Zador. "Sound Processing Takes Motor Control." *Nature*, vol. 513, 2014, pp. 180–81.

Lowlands by Susan Philipsz. 47 Film, 2010. *YouTube*, www.youtube.com/watch?v=UWeKzTDi-OA.

McCarthy, John. "How the Brain Does Attention Is Still Unknown." *Scientific American*, 10 Oct. 2012, blogs.scientificamerican.com/ guest-blog/how-the-brain-does-attention-is-still-unknown/.

Minor, Dale. "The Second Battle of Selma." Narrated by Minor. *From the Vault: Pacifica Radio Archives*, 15 Mar. 2013, fromthevaultradio.org/home/2013/03/15/ftv-357-the-second-battle-of-selma/.

Morley, Madeleine. "The Unheard Symphony of the Planet." *The New York Times*, 3 Jan. 2023, nytimes.com/2023/01/03/arts/seismology-raspberry-shake-earth.html

Motter, Bucky. Guitar Pull Community Event. Atlanta, GA, 09 June 2019.

Oliveros, Paulina. *Deep Listening: A Composer's Sound Practice.* iUniverse, Inc., 2005.

Perryman-Clark, Staci M. "The Promises and Perils of Higher Education: Our Discipline's Commitment to Diversity, Equity, and Linguistic Justice." *2022 Call for Proposals*, Conference on College Composition and Communication, NCTE, ccc.ncte.org/cccc/call-2022.

Pollock, David. "Susan Philipsz Sounds the One O'Clock Gun at 2012 Edinburgh Art Festival." *The List*, 11 July 2012, edinburghfestival.list.co.uk/article/ 43216-susan-philipsz-sounds-the-one-oclock-gun-at-2012-edinburgh-art-festival/.

Powell, Malea, et al. "Our Story Begins Here: Constellating Cultural Rhetorics," *enculturation*, 18, 2014, enculturation.net/our-story-begins-here.

Program for "Re/Call: A Curated Exhibit and Community Performance Event." 1–3 Sept. 2018,

RedLine Contemporary Art Center, Rocky Mountain Land Library, A.P. Fiedler, 2018.

Racing to Extinction. Directed by Louis Psihoyos, Discovery Channel, 2015.

Ratcliffe, Krista. *Rhetorical Listening: Identification, Gender, Whiteness.* Southern Illinois UP, 2005.

"Re/Call: A Curated Exhibit and Community Performance Event." 1–3 Sept. 2018, RedLine Contemporary Art Center, Rocky Mountain Land Library.

Reid-Musson, Emily. "Intersectional Rhythmanalysis: Power, Rhythm, and Everyday Life." *Progress in Human Geography*, vol. 42, no. 6, 2018, pp. 881–97.

Revol, Claire. "Henri Lefebvre's *Rhythmanalysis* as a Form of Urban Poetics." *The Routledge Handbook of Henri Lefebvre, The City and Urban Society*, edited by Michael E. Leary-Owhin and John P. McCarthy, Routledge, 2019, pp. 173–82.

Rice, Jeff. "Beneath the Tree." 2022. *Ecosystem Sound*, ecosystemsound.com/beneath-the-tree.

Rice, Jeff. "The Making of Ka-knowledge: Digital Aurality." *Computers and Composition*, vol. 23, nos. 1–2, 2006, pp. 66–279.

Rickert, Thomas. *Ambient Rhetoric: The Attunements of Rhetorical Being.* U of Pittsburgh P, 2013.

Rickert, Thomas, ed. "Writing/Music/Culture." Special issue of *enculturation: A Journal of Rhetoric, Writing and Culture*, vol. 2, no. 2, 1999, enculturation.net/2_2/index.html.

Rivers, Nathaniel A. "Geocomposition in Public Rhetoric and Writing Pedagogy." *College Composition and Communication*, vol. 67, no. 4, 2016, pp. 576–606.

Roberts, Dmae. "*Mei Mei*, A Daughter's Song." Narrated by Roberts. *Public Radio Exchange*, 17 Mar. 2006, beta.prx.org/stories/7184.

Robinson, Dylan. *Hungry Listening: Resonant Theory for Indigenous Sound Studies.* U of Minnesota P, 2020.

Rocky Mountain Land Library: A Resource Linking Land and Community. Rocky Mountain Land Library, landlibrary.wordpress.com.

Rosser, Chris. Scheduled Studio Session with Mary Hocks, Hollow Reed Studio, Asheville, NC, 18 July 2019. https://www.chrisrosser.com/studio.

Sansbury, Matthew. "It Better Get Better: A 'No H8' Space That's Bully-free." *Wordpress*, 2012, itbettergetbetter.wordpress.com/.

Sayers, Jentery. "Audiography," 2010, jenteryteachers.com.

—. *How Text Lost Its Source: Magnetic Recording Cultures.* 2011. U of Washington, PhD dissertation.

—. "Audio Culture Series: Scaffolding a Sequence of Assignments." *Sounding Out*!, 3 Sept. 2012, soundstudiesblog.com/author/jentery/.

Schaeffer, Pierre. "Etude aux Chemin de Fer." 1948. *YouTube*, youtube.com/watch?v=N9pOq8u6-bA.

Schafer, R. Murray. *The Tuning of the World.* McClelland & Stewart, 1977.

Schmutz, Vaughn and Alison Faupel. "Gender Consecration in Popular Music." *Social Forces* vol. 89, no. 2, 2010. 685–707.

Scott, Sarah Wallace. *Four Songlines.* 2018, RedLine, Rocky Mountain Land Library.

Selfe, Cynthia L. "The Movement of Air, the Breath of Meaning: Aurality and Multimodal Composing." *College Composition and Communication*, vol. 60, no. 4, 2009, pp. 616–63.

Shively, Kim. "Recalling Re/Call." 2019. Rocky Mountain Land Library Website, landlibrary.wordpress.com/2019/01/09/recalling-re-call/

Expo: Sick Architecture. CIVA (Centre for Information, Documentation and Exhibitions), 2022, civa.brussels/en/exhibitions-events/expo-sick-architecture.

Song Circle, Atlanta, GA, 27 Oct. 2019.

Song Circle, Asheville, NC, 19 June 2021.

Stacy, Rosie, et al. "Singing for Health: An Exploration of the Issues." *Health Education*, vol.102, no. 4, 2002, pp. 156–62.

Stedman, Kyle. *Musical Rhetoric and Sonic Composing Processes.* 2012. U of South Florida, PhD dissertation.

Stedman, Kyle D., et al., eds. *Tuning in to Soundwriting.* Intermezzo, 2021, www.intermezzo.enculturation.net/14-stedman-et-al/index.html.

Sterne, Jonathan. *The Audible Past: Cultural Origins of Sound Reproduction.* Duke UP, 2003.

—. "Sonic Imaginations." *The Sound Studies Reader*, edited by Jonathan Sterne, Routledge, 2012, pp. 1–20.

Stone, Jonathan. *Listening to the Lomax Archive: The Sonic Rhetorics of African American Folksong in the 1930s*, U of Michigan P, 2021, press.umich.edu/Books/L/Listening-to-the-Lomax-Archive

—. "Listening to the Sonic Archive: Rhetoric, Representation, and Race in the Lomax Prison Recordings." *enculturation: A Journal of Rhetoric, Writing and Culture*, 19: (May 2015).

—. *Listening to the Lomax Archive: The Sonic Rhetorics of American Vernacular Music in the 1930s.* 2015. U of Illinois, PhD dissertation.

Stone, Jonathan, and Steph Ceraso (Eds). "Sonic Rhetorics." Special issue of *Harlot of the Arts*, no. 9, 2013, harlotofthearts.org/index.php/harlot/issue/view/9.

Stoever, Jennifer Lynn. *The Sonic Color Line: Race and the Cultural Politics of Listening.* New York UP, 2016.

Sublet, Anna. "The Healing Power of Singing." *The New York Times*, 27 Oct. 2020, https://www.nytimes.com/2020/10/27/well/mind/choir-singing-music-anxiety-stress.html

Tepp, Tory. *The Forge.* 2018, RedLine, Rocky Mountain Land Library.

Thulin, Samuel. "Sound Maps Matter: Expanding Cartophony." *Social & Cultural Geography*, vol. 19, no. 2, 2018, pp. 192–210. 2018.

Tsing, Anna Lowenhaupt. *The Mushroom at the End of the World: On the Possibility of Life in Capitalist Ruins.* Princeton UP, 2017.

Understanding Sound. National Park Service, 3 July 2018, nps.gov/subjects/sound/understandingsound.htm.

Valenzuela, Cecilia Angélica and Magnolia Landa-Posas. "AudioVoice: A Relational, Subaltern Praxis of Listening to Testimonios and Composing with Sound." *Kairos: Rhetoric, Technology, Pedagogy*, vol. 26, no. 1, 2021, https://kairos.technorhetoric.net/26.1/topoi/aguilar-et-al/cecilia-valenzuela.html.

VanKooten, Crystal. "Singer, Writer: A Choric Exploration of Sound and Writing." *Kairos:A Journal of Rhetoric, Technology, Pedagogy*, vol. 2, no. 1, 2016, kairos.technorhetoric.net/21.1/inventio/vankooten/chora.html.

—. *Developing Meta-awareness about Composition through New Media in the First-year Writing Classroom.* 2014. U of Michigan, PhD dissertation.

Voegelin, Salome. *Listening to Noise and Silence: Towards a Philosophy of Sound Art.* Continuum International Publishing Group, 2010.

Westerkamp, Hildegard. "Kits Beach Soundwalk." 1989. *YouTube*, youtube.com/watch?v=hg96nU6ltLk.

Whyte, Kyle Powys. "Time as Kinship." *The Cambridge Companion to the Environmental Humanities,* edited by Jeffrey Cohen and Stephanie Foote, Cambridge UP, 2021, pp. 39–55.

Wishart, Trevor. "Tongues of Fire." 1994. *YouTube*, youtube.com/watch?v=Ude4717dlsQ.

Wolf, Jeremy. *Anthropocene Symphonia.* 2018, RedLine, Rocky Mountain Land Library.

Wysocki, Anne Frances, et al. *Writing New Media: Theory and Applications for Expanding the Teaching of Composition*, Utah State UP, 2004.

Yancey, Kathleen. "Made Not Only in Words: Composition in a New Key." *College Composition and Communication*, vol. 56, no. 2, 2004, pp. 297–328.

Yergeau, M. Remi. "Multimodality in Motion: Disability and Kairotic Spaces." Special issue of *Kairos: A Journal of Rhetoric, Technology, and Pedagogy*, vol. 18, no. 1, 2013, kairos.technorhetoric.net/18.1/coverweb/yergeau-et-al/index.html.

Index

About the Authors

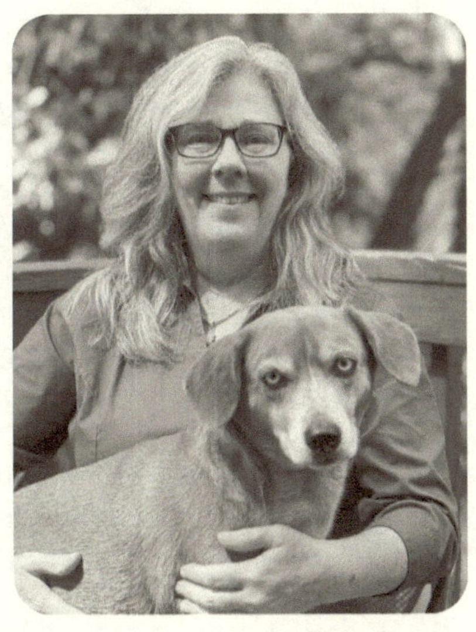

Mary Hocks is Associate Professor of English at Georgia State University in downtown Atlanta, where she teaches undergraduate and graduate courses in composition theory and practice, contemporary rhetorical theory, and digital writing. She is the former director of the university writing studio and the founding director of writing across the curriculum. Previously, she was director of the comprehensive writing Program at Spelman College. Her research has appeared in *College Composition and Communication, Writing Program Administration, Rhetoric Review*, and *Computers and Composition*, among others. Her co-edited collection, *Eloquent Images: Word and Image in the Age of New Media*, appeared from The MIT Press. Hocks regularly composes and performs original music, and she draws upon her experiences as a singer-songwriter and collaborator in her teaching and research. Her album *Spinning the Sky*, with thirteen songs featuring regional musicians, was released in July 2008.

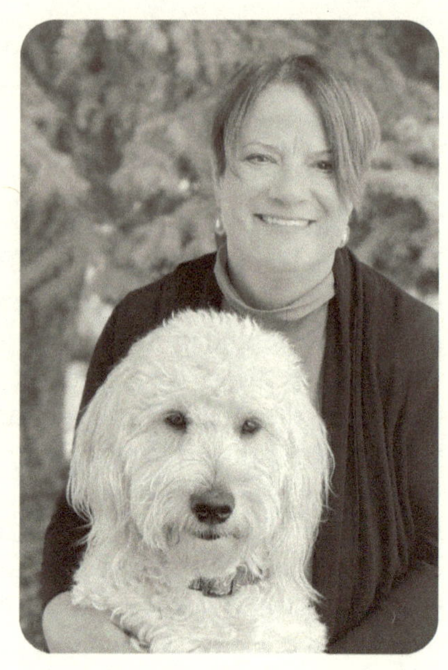

Michelle Comstock is an associate professor of Writing, Rhetoric, and Technology in the Department of English at the University of Colorado Denver. Her research, which is informed by scholarship in sonic rhetoric, cultural studies, and the environmental humanities, has appeared in *Computers and Composition, Community Literacy, Rhetoric Review*, and *JAC*, among others. She co-authored *Composing Public Space* (Heinemann) with Lil Brannon and Mary Ann Cain and is currently co-producing a grant-sponsored digital audio tour of the Auraria Campus and its history of displacement. She regularly teaches courses on multimedia production and serves as the faculty coordinator and instructor for the department podcast *CultureKlatsch*.

www.ingramcontent.com/pod-product-compliance
Lightning Source LLC
LaVergne TN
LVHW051005080826
845145LV00009B/2469

* 9 7 8 1 6 4 3 1 7 4 5 4 9 *